T0332170

Cultural Evolution

In this book, Kate Distin proposes a theory of cultural evolution and shows how it can help us understand the origin and development of human culture. Distin introduces the concept that humans share information not only in natural languages, which are spoken or signed, but also in artefactual languages like writing and musical notation, which use media that are made by humans. Languages enable humans to receive and transmit variations in cultural information and resources. In this way, they provide the mechanism for cultural evolution. The human capacity for metarepresentation – thinking about how we think – accelerates cultural evolution, because it frees cultural information from the conceptual limitations of each individual language. Distin shows how the concept of cultural evolution outlined in this book can help us understand the complexity and diversity of human culture, relating her theory to a range of subjects including economics, linguistics and developmental biology.

Kate Distin was educated at Cambridge University and the University of Sheffield. She is the author of *The Selfish Meme: A Critical Reassessment* (Cambridge, 2005) and the editor of the award-winning *Gifted Children: A Guide for Parents and Professionals* (2006).

Cultural Evolution

KATE DISTIN
Independent Scholar

CAMBRIDGE
UNIVERSITY PRESS

CAMBRIDGE
UNIVERSITY PRESS

32 Avenue of the Americas, New York NY 10013-2473, USA

Cambridge University Press is part of the University of Cambridge.

It furthers the University's mission by disseminating knowledge in the pursuit of education, learning and research at the highest international levels of excellence.

www.cambridge.org
Information on this title: www.cambridge.org/9780521769013

First published 2011

A catalogue record for this publication is available from the British Library

Library of Congress Cataloguing in Publication data

Distin, Kate, 1970–
Cultural evolution / Kate Distin.
p.   cm.
Includes bibliographical references and index.
ISBN 978-0-521-76901-3 (hardback) – ISBN 978-0-521-18971-2 (pbk.)
1. Human evolution.   2. Social evolution.   3. Language and languages – Origin.
4. Human beings – Origin.   I. Title.
GN281.D58   2010
303.4–dc22   2009051653

ISBN 978-0-521-76901-3 Hardback

Contents

1. Introduction: "Small Consequences of One General Law"   *page* 1

PART I. THE INHERITANCE OF CULTURAL INFORMATION

2. What Is Information?   11

3. How Is Information Inherited?   25

PART II. THE INHERITANCE OF CULTURAL INFORMATION:
NATURAL LANGUAGE

4. Natural Language and Culture: The Biological
Building Blocks   49

5. How Did Natural Language Evolve?   62

6. Language, Thought and Culture   78

PART III. THE INHERITANCE OF CULTURAL INFORMATION:
ARTEFACTUAL LANGUAGE

7. How Did Artefactual Language Evolve?   89

8. Artefactual Language, Representation and Culture   107

9. Money: An Artefactual Language   126

10. Money: The Explanatory Power of Artefactual Languages   146

PART IV. THE RECEIVERS OF CULTURAL INFORMATION

11. How Does Human Diversity Affect Cultural Evolution?   169

PART V. THE EXPRESSION OF CULTURAL INFORMATION

12. Aspects of the Cultural Ecology   185

13. Patterns of Cultural Taxonomy   200

14. Conclusion: A Representational Understanding of Cultural
    Evolution   220

Appendix: What about Memetics?   231
Acknowledgements   235
Bibliography   239
Index   263

Cultural Evolution

1

Introduction

"Small Consequences of One General Law"

Bachan was born into a world in which his mother's salwar kameez and his father's turban, his sister's favourite pakora and his brother's bhangra collection already existed. His parents are British Asians, and he inherits a culture as different from his grandparents' native Punjabi culture as it is from his best friend's native English culture. His grandparents had no electricity in their homes when they were his age. The mobile telephone did not exist when his parents were born. The Harry Potter books had not been written when his sister was born. Bachan's world includes the Fairtrade Foundation and reality television, ongoing debates about bioethics and the environment, and a previously unimaginable range of purchasing options for every type of product, from snacks to satellite navigation systems. His world is bounded by the assumptions and rules that his parents make on his behalf: he absorbs, without even noticing what is happening, his parents' behavioural standards, moral judgements, religious beliefs and practices, educational values and parenting methods. He speaks English with the characteristic Midlands Asian accent, but he can only understand Punjabi, not speak the language. He wears his hair uncut, covered by a patka, but he attends an Anglican primary school. Would he pass Norman Tebbit's cricket test? Probably not. Nor would I, for that matter, if you asked me whether I'd support Yorkshire in a cricket match against the county where I now live. Like Bachan, however, I consider an emotional attachment to the place of my family's origin irrelevant to questions about where I make my home, how integrated I am into the local community, or indeed what I can give to that community.

Our genes are fixed. We cannot choose to inherit one parent's perfect eyesight in preference to the other's astigmatism, or to be taller than we are. This is not to say that the same genes are always expressed in identical ways: nutritional variation will affect a child's height, and educational variation will affect how she performs in an IQ test; identical twins are not identical people – but nor can they choose to alter their genes, in the way that they can choose to alter their appearances or to develop different skills, to study different subjects or to practice different religions. We can make choices about our culture in a way that we simply cannot about our nature. We can choose, for example, how we respond to our genetic inheritance. We cannot choose not to have been born with poor eyesight, but if we live in a country where the options are available then we can choose whether to wear glasses or contact lenses to correct our vision, or even to opt for surgery. We can also choose how we respond to our cultural inheritance. We cannot choose not to have been born into a family that practices a particular religion, speaks a particular language or takes a particular view of education, but we can choose whether to adopt that religion ourselves, whether to learn to speak a different language, and whether we agree with that view of education.

Crucially, our responses to our cultural inheritance will also be passed on to our children, in a way that our responses to our genetic inheritance will not. Eye surgery does not alter a person's genes: her children stand as great a chance of inheriting her poor eyesight if they are conceived after she had surgery as they do if they are conceived before it. If she rejects her parents' atheism, however, and becomes a Christian, then her children will be brought up not as atheists but as Christians. If she regrets the fact that her parents did not put a high value on education, so that she left school barely literate and numerate, then she might make an effort to improve her own skills and will also pass on to her children an entirely different attitude to school from the one that her parents gave her.

Culture of this kind is unique to the human species. Debates continue about the extent to which nonhuman species, the primates in particular, are capable of using language or tools, of learning from one another and possessing what might be called a culture, but incontrovertibly no other species has developed anything like the depth and breadth of human culture. The inherited tendencies of many species can be given direction by exposure to an environmental stimulus: newly hatched goslings or ducklings, for example, have an inherited tendency to become socially bonded to the first moving object they encounter, and this tendency is directed by whichever moving object happens to provide the right environmental stimulus. As well as the information that they carry in their genes, which gives rise to these inherited tendencies, many species also carry information in their brains: they are able to make and remember associations between events, and their behaviour is subsequently influenced by these associations as well as by the environment and their biology, as when Pavlov's famous dogs drooled at the ringing of a bell because they had learned to associate its sound with the arrival of food. In addition to information like this, which they have learned by themselves, members of some species can imitate the behaviour of conspecifics: the behaviour of these creatures is influenced by their genes and by their environment, by the information in their own brains and also by the behaviour of other creatures. Members of only one species, so far as we can tell, carry information around externally as well as in their brains and in their genes. Human behaviour is influenced by our genes and by our environment, by the information in our own brains, by the behaviour of others and – uniquely – by the information in conspecifics' brains, which we can access via their speech and via written symbols. We are alone in having set information free from the confines of genes and brains, and the result is human culture.

How has what we humans learn from one another become so much more complex and diverse than what members of any other species learn from one another? It will be apparent that genetics can offer only partial answers to questions about culture. Our capacities for learning, and for putting into practice what we learn, are shaped by our genes; the ways in which we respond to what we learn will often be influenced by our genes; but vast swathes of human culture are immune to the explanatory power of genes – if for no other reason than that they change much too quickly for our genes to keep up with them. Human culture develops so rapidly and radically that it can be hard for us to see what is happening. It can be difficult enough to keep up with the impact of its changes on our everyday lives, never mind to reflect on them and analyse their structure. Nor is it correct to suppose that just because a particular trait is biologically adaptive for humans it will therefore be genetic rather than cultural in origin: it would be ridiculous to suggest that the adaptive advantages of modern inventions like computers and passenger airbags are the product of modern changes in the human genetic code. This means that Daniel Dennett's famous question, "Cui bono?" (1995: 325), will not always be a reliable guide to behavioural origins: culturally inherited behaviour can be biologically adaptive even though it is not the product of biological evolution. The situation is further complicated by the fact that humans are able to choose, in at least some circumstances, whether to heed the promptings of our genes or to serve our own self-interests as individual human beings. In a Darwinian creature like a bee, there is no conflict between its own goals and those of its genes: the bee's goals simply are its genes' goals, and it will sacrifice itself to those goals if necessary (Stanovich and West 2003). Humans, on the other hand, are uniquely capable of putting our own individual interests above those of our genes: many people choose to use contraception, for example, and members of the emergency services regularly put their own lives in danger for the sake of strangers.

Given both how much more of human behaviour than of any other creature's is learned rather than inherited, and the immense complexity and variety of human behaviour, it would be surprising if it were a straightforward matter to separate the biological from the cultural influences on our lives. The decision will be more difficult to make about some traits than about others. Yet evolutionary theory offers a methodology for studying the patterns of cultural change, in the same way that it has unified our understanding of biology. Charles Darwin (1859: 263) described inherited instincts as "small consequences of one general law," which he summarised as "multiply, vary, let the strongest live and the weakest die." The theory of cultural evolution sees each element of culture's complex diversity as another small consequence of the same general law operating in a different jurisdiction: the realm of culture. It is not uncommon to hear about the "evolution" of a car design, religious doctrine or recipe, and there is a consensus among many researchers that we can take this talk literally (e.g., Blute 2007; Deacon 1999; Dennett 2006b; Henrich, Boyd and Richerson 2008; Marsden 1998; Mesoudi 2007a). The theory of cultural evolution contends that the changes and developments in all areas of human culture can truly be said to evolve: that they, just as much as the changes and developments in nature, can be described by an evolutionary algorithm; and that a convincing theory of cultural evolution can play the same unifying role across the social, psychological and behavioural sciences as evolutionary theory has played in biology.

Towards a Better Understanding of Culture

To what extent is cultural evolution genuinely analogous to the more familiar biological evolution, which all life on earth has in common? Strictly speaking, cultural evolution is a different example of the same type of process as neo-Darwinism rather than a simple analogue of it. Our familiarity with biology can therefore be exploited as a resource for our growing understanding of culture, but we should not expect the particular details of biological evolution to carry over into cultural evolution.

Evolution is a gradual, intergenerational process of change in a population's characteristics, and cannot happen unless variations in that population's characteristics are inherited across many generations. Our knowledge of biology indicates that we can usefully describe this process in terms of the *information* that inheritance mechanisms make available to each generation. In the opening chapters of this book I develop a theory of information and its inheritance, which not only brings all aspects of biological and cultural information under one explanatory umbrella, but also enables us to understand how cultural evolution has taken off in humans as a process independent from biological evolution. My conclusion is that information can never exist in isolation, but must always be transmitted to a receiver *that can interpret it and respond appropriately.* Information is any variation that a receiver discretely represents, and it can only be acquired from a representational source if the receiver discretizes it in the same way that the source does. This means that evolution depends on each generation's ability to interpret and express the information that it inherits. Genetic variants, for example, rely for their interpretation and expression on the next generation's cytoplasmic inheritance of the cellular transcription and translation machinery. Viral DNA achieves the same aim by hijacking that machinery in organisms. Variations in cultural information rely for their inheritance and expression on the existence of receivers who understand the particular system in which they are represented.

In the light of this theory of heritable information, the significance of language for cultural evolution becomes apparent: the emergence of natural language created generation upon generation of receivers for the information that it represents, ensuring the persistent heritability on which evolution depends. This theory also explains why cultural evolution is unique to our species: only humans have developed a system that ensures the persistent heritability of both cultural information and its means of interpretation and expression. Why this should be the case, and how natural language evolved, are the subjects of Chapters 4 and 5.

Natural language is an immensely powerful tool of communication, but by itself it cannot account for the nature of the changes in human culture over the millennia since it first emerged. I describe how the result of natural language evolution was an explosion in the amount of information that early humans were able to trade – and there is only so much information that we can hold and manage in our brains alone; even in our collective brains. There came a point at which there was too much cultural information available for individuals to manage reliably using memory and speech alone. When cultural information had expanded beyond the point that the brain could manage independently, humans began to develop artefactual symbols in order to store and manipulate this excess cultural information. The result is that cultural information is represented not only in natural languages like English, Mandarin and Nicaraguan Sign Language, which use the biologically evolved media of human vocal chords and gestures, but also in what I shall call artefactual languages, like the written word or mathematical notation, the conventions of cartography or the vocabulary and formatting requirements of computer programming languages: languages that use the culturally evolved media of artefacts like paper and ink or keyboard and text editor.

Evidence from fields as diverse as archaeology and economic development, engineering and music, psychology and the history of technology, all points to the role of artefactual languages in providing substrates for the cultural evolution of information that natural language cannot contain and human brains cannot manage without support. In the competition between cultural information, artefactual representational methods had advantages like stability, capacity and accuracy. It is not until we acknowledge the role of artefactual as well as natural languages in cultural evolution that we can really begin to explain how cultural complexity is maintained and transmitted.

The advantage of my approach to this subject is that it enables the evolution of human culture to be explained as a token of the same type of process as the evolution of the natural world, and simultaneously makes clear why it is unique: human culture, which has its origins in information that is represented in natural and artefactual languages, cannot be shared by receivers that are not biologically prepared to learn those languages. Perhaps the most significant aspect of human preparedness[1] for language acquisition is our capacity for metarepresentation: thinking not only about the content of our representations but also about the representations themselves, lifting information out of its original context, reflecting on it in the abstract, choosing how best to represent it and in what medium. Languages do not only act as receptacles and conduits for

---

[1] Evolution, needless to say, has no foresight and is not in the business of preparing species for anything, but nonetheless, there came a time when our ancestors were both language ready and culture ready, as a result of prior evolutionary processes.

information; they also play a crucial role in shaping our cognition, by restricting us to a particular way of thinking about the particular information that each is capable of carrying. Metarepresentation enables us to escape the conceptual limitations of a particular language, freeing us from its cognitive constraints so that we can re-represent the information that it carries, or recombine it with information from another source.

Evidence from research into giftedness and rationality indicates that there is a spectrum of innate metarepresentational ability in humans, which can also be affected by educational levels. This diversity among the human receivers of cultural information is bound to affect the course of cultural evolution, and in particular it will help to shape the patterns of cultural taxonomy. Drawing on lessons that have been learned only recently in the fields of prokaryotic and viral taxonomy, I show how cultural variations are sometimes subject to the restrictions of species-like barriers; how the porousness of these barriers varies; and how cultural taxonomy, as a consequence, will often need to chart reticulate relationships between polythetically defined classes of cultural artefacts and behaviours. Nonetheless, evidence from a range of studies has shown that in a surprising number of cases cultural phylogenies are tree-like, providing further support for the thesis that there are some species-like barriers in culture.

An understanding of the relationship between cultural information and human agents, of metarepresentation, and of the distinction between communicative and representational uses of language, casts new light on a range of cultural phenomena. In my exploration of the evolution and modern use of money, I draw together evidence from archaeology, economics and (rather surprisingly) the Eurovision Song Contest, for the claim that money can helpfully be seen as an artefactual language in which information about value is represented and exchanged, and I show how this representational view of money can explain a host of monetary phenomena.

The concluding chapter submits that a representational theory of cultural evolution has a broader application and a greater explanatory value than is often acknowledged. Indeed, perhaps it can even account for the varying reactions that readers will have to this book. If information is, as I shall argue, any variation that a receiver can discretely represent, then the heritability of information is crucially dependent on receivers. If writing is, as I shall argue, an artefactual language for the representation of cultural information, then the heritability of that information is crucially dependent on its readers. All writing carries information, but the precise content of that information will be shaped by each individual reader. The message that you take from this book will not be identical to the message that any other reader takes from it. You exert a unique force on the currents of cultural evolution.

PART I

THE INHERITANCE OF CULTURAL INFORMATION

2

What Is Information?

At the heart of my theory of cultural evolution is the claim that human culture, as much as our nature, is the product of evolving information. Culture is the behavioural and artefactual product of interactions between humans and cultural information, and cultural evolution is the product of heritable variations in that information. This means that our understanding of cultural evolution must be founded on an understanding of heritable information.

Informational language provides biologists with a handy conceptual tool, particularly when thinking about genes: it is common to hear talk of genetic *information* being *transcribed, translated, edited, expressed* and *transmitted* from one generation to the next. It is a little surprising, then, to find that there is no clear consensus about what information is, exactly. If evolution can really be understood as the product of heritable information, then there seems to be a major conceptual omission at the heart of our understanding of evolutionary theory.

In this chapter and the next, I develop a theory of heritable information that is broad enough to encompass biological as well as cultural information but detailed enough to provide answers to individual questions in both fields. It enables us to distinguish, for example, between inheritance mechanisms that can contribute to long-term evolution and those that cannot. It helps to resolve debates about whether genes carry a causally special type of information in nature, and why human culture is so uniquely complex and extensive. It brightens the light that earlier information theories have shed on information's receivers, and it manifests a range of empirical predictions about what happens to information if we keep the source fixed but vary the receiver, and about which sorts of receivers will be able to acquire and transmit which types of information.

Information Theory, Biology and Culture

Information theory has its origins in a mathematical model of communication that was developed by Claude Shannon and popularized by Warren Weaver in the 1940s (Weaver and Shannon 1949). The Shannon-Weaver model represents communication as a process in which an information source selects a message, which a transmitter then encodes into a signal to be sent over a communication channel, to a receiver that decodes the signal back into the original message, handing it over to the destination. In speech, for example, the transmitter is the speaker's vocal equipment, the signal consists of sound waves, the hearer's ear is the receiver, and the speaker and hearer themselves are, respectively, the source of and destination for the information.

A simple Shannon-Weaver-type model of communication enables us to express our understanding of inheritance in informational terms: information is inherited when one generation acts as a source that transmits a signal to the following generation, which reacts to the signal and interprets it. This is the model that has been adopted by Eva Jablonka and Marion Lamb (2005), who characterize biological evolution as the product of four types of inheritance mechanism, each of which transmits information or resources from one generation (the source) to the next (the receiver). Although an information source itself does not usually change when a receiver acquires information from it (2005: 109), the receiver's functional state will be changed in response to the source's form and organization (2005: 54). For instance,

> a train timetable is something that can affect the potential actions of the person who reads it; a recipe for an apple pie can affect the baking activities of a cook; the length of daylight can affect the flowering time of a plant; an alarm call can affect the behaviour of the animal that hears it; a DNA sequence can affect the phenotype of the organism. In all cases, a receiver can react to the source in a functional way that corresponds to the source's particular form. When the receivers react in such a way, they are interpreting the source's organization. (2005: 54)

The emphasis that Jablonka and Lamb place on the role of receivers in the transmission of information will prove crucial for a convincing account of cultural evolution. Following the Shannon-Weaver model, the authors describe clouds, train timetables and DNA, for example, as sources of information *because* all have receivers that can react to them and interpret them (2005: 109). What Jablonka and Lamb do not fully explain is *how* a particular receiver is able to react appropriately to the form of an information source. In fact, this is a question that Shannon-Weaver-type models fail to address, because they are more concerned with information's transmission than they are with its interpretation. The problem with employing a "postal metaphor" (Chandler 1994) of communication, in which the source sends a package of information to a receiver, is that it encourages us to think of the source as actively determining the meaning of the message, and of the receiver as a passive target. In reality, of course, there are multiple ways in which a receiver can interpret information, and decoding will not necessarily be a mirror image of encoding (Chandler 1994). The receiver plays a much more active role in the communication of information than is sometimes acknowledged, and an improved theory of heritable information will rest on a better understanding of the ways in which receivers interpret and respond to information. This chapter explores a variety of types of information and asks what it is, in each case, that enables a receiver to interpret and respond appropriately to variations in the information source.

Associative Learning and Innate Knowledge

"We say that a cloud, something physical, conveys information about the weather; a clock provides information about the time; the smell in a restaurant carries information about the food; newspapers contain information about world events. Biologists would say that a DNA sequence carries information about the sequence of amino acids in a protein, or where a regulator binds; they would also say that a bird's song carries information about its species, and a mother's playful behavior carries information about the world for her child" (Jablonka and Lamb 2005: 109).

Let's look, for example, at the information that the smell in a restaurant carries about the restaurant's food. If a customer has learned from past experience to associate a particular smell with a particular food, then in the future when he experiences the same smell he will be able to identify the food from which it comes. He plays the role, when this happens, of a receiver who can react to the smell in a way that corresponds to its particular form. So the nature of his reaction – in this case, his correct identification of the food – is determined not only by the smell itself but also by his past associative learning experience.

Similarly, when we say that clouds "convey information about" the weather, what we mean is that we have learned to recognize a predictive correlation between the two, such that when we see clouds, we can make predictions about other aspects of the weather. *Weather* is a broad term for the short-term state of the atmosphere, which encompasses precipitation, temperature, air pressure, humidity, sunshine and wind velocity, as well as cloud cover. Just as a smell does not contain information about the food to anyone who has not already experienced the association between that food and its smell, so a cloud carries no information to people who have not learned to associate that type of cloud with other meteorological events.

A similar learning experience is necessary before a bird's song can bring us information about its species. Although I can identify many British garden birds from their appearance, the vast majority of their songs bring me no information at all – and for some people, even the birds' appearance carries no information. Like the smell of food and the appearance of clouds, the sound of a bird's song is a sensation that we can learn to associate with a particular entity: a way in which we can detect something's presence at a distance. But it can carry information over that distance only if we have first learned to associate the sensation with its origin. In all of these examples, then, the information that a receiver acquires from a source is dependent on a prior learning experience.

Appropriate responses to information include more than mere iden-tification of its source, however. Vervet monkeys, for instance, have an appropriate behavioural response to each of the three distinct noises that members of their species make on noticing a leopard, an eagle or a snake: hiding under bushes in response to the eagle call, running up a tree in response to the leopard call, and looking at the ground while standing on tiptoe in response to the snake call (Cheney and Seyfarth 1990). What information do these alarm calls carry, and how do the monkeys acquire it? It is unclear, at first glance, whether the calls carry information about the action that the hearers should take or about the predator to which that action would be appropriate. But the evidence, for Diana monkeys at least (Zuberbühler, Cheney and Seyfarth 1999), is that monkeys that notice an eagle five minutes after hearing an eagle alarm call show less alarm than they do if they notice a leopard after hearing the eagle alarm call, and this leads us to conclude "that the alarm calls do not merely trigger the relevant evasive action, with no representation of the specific source of danger being kept in the head: the Diana monkeys, on hearing a leopard alarm call, keep the idea of a leopard in their minds for at least five minutes; and likewise with the eagle alarm call" (Hurford 2007: 227). So the alarm call affects not only the animal's actions but also its mental representations.

In the earlier examples that we considered, receivers relied on an associative learning experience to acquire information from a source. Learning also plays a part in the monkeys' interpretation of alarm calls, but it is not the only factor: both the sending and the receiving of alarm calls appear to be under broad genetic control, fine-tuned by learning experience. For instance, very young vervets reserve leopard alarm calls primarily for terrestrial mammals and eagle alarm calls for birds, but adult vervets give alarm calls for a much narrower range of creatures, which suggests that they have learned to distinguish between species as well as between broad classes of predator (Hurford 2007: 230). Infant vervet responses to alarm calls are also more generalized than adult responses, and more dependent on the promptings of visible conspecifics. The implication is that these monkeys are born with an innate knowledge of the association between the different alarm calls and the broad classes of predators that those calls are about, and that, in addition, they are able to learn to associate the calls with more specific groups of predators. Creatures that are unaware of such an association are unable to act on the alarm calls.

In all of the examples that we have considered so far, the receiver has been a creature with some measure of awareness and learning ability. As Jablonka and Lamb (2005) point out in some of their other examples, however, information can also be transmitted to receivers of a very different kind. What is going on, for instance, when flowering plants respond to the length of daylight? Plants' response to the relative lengths of day and night – their photoperiodism – is controlled by changes in a light-sensitive pigment-protein complex called phytochrome. Phytochrome exists in two interchangeable forms: in daylight it is converted to an active form, and during the night it reverts to an inactive form. When the active form reaches a particular level, which differs between plants, it triggers a variety of changes in the plant, including flowering. Each type of plant requires a particular period of darkness to convert the active form of phytochrome back to the requisite level of the inactive form.

The plant, in other words, has a chemical switch that links night length to flowering time. Just as vervet monkeys have an innate knowledge of the association between conspecifics' alarm calls and broad classes of predator, so the plant has an innate "knowledge" of the association between night length and flowering time. If it did not, then it could not react to night length as it does. For a plant to flower at the usual time for its species, it is not sufficient for the night to be a particular length; the plant needs also to have the appropriate light-responsive mechanism to react to that night length. Its reaction relies on the existence of a link between a particular night length and the processes that lead to flowering – and the nature of that link determines the way in which the plant responds to night length.

This point is most clearly illustrated in a different example (not one of Jablonka and Lamb's): the rotary switches that can still be found in some electromechanical central-heating time controls. These programmers have a dial, which acts as a clock. The dial is marked from zero to twenty-four, and as it rotates the numbers pass a marker whose position indicates the current time. In addition to this fixed marker, there is also a pair of movable tappets, which act as on or off switches. One tappet can be positioned to the time that the heating is to be switched on and the other to the time that it is to be switched off; and in fact there is usually more than one pair, so that the heating can be switched on for more than one period of each day.

This programmer clock (the source) carries information to the central-heating system (the receiver) about the time of day, just as night length (the source) carries information to a plant (the receiver) about the time of year. The plant reacts to its information by flowering at the right time, and the central-heating system reacts to its information by switching on or off at the right time. In the plant's case, the right time has been determined by evolutionary history, whereas in the programmer's case it has been set by human intention, but these histories are irrelevant to the processes that are at work. In each case, the reaction could not happen in the absence of the relevant switch: an evolved, chemical switch in the plant and a man-made, mechanical switch in the programmer. In both cases, the switch enables the receiver to react appropriately to the information. In neither case would the receiver be able to take any information from the source in the absence of the switch: night length could vary and the clock dial turn forever, but without a chemical switch or tappet they would not produce any reaction at all.

Discrete Representational Knowledge

If innate connections explain how night length conveys meaningful information to a plant or alarm calls to a monkey, and learned connections explain how clouds and smells and birds' songs can convey meaningful information to a human, then we can begin to see a pattern emerging. What all of these examples have in common is a one-to-one connection, or at the most a very limited repertoire of links, between the informational input and the receiver's reaction.

In the case of a more complex source, like the clocks, recipes and train timetables that Jablonka and Lamb mention, a different sort of connection needs to exist between the information source and the receiver's reaction, because these sources carry much more information than an isolated smell or bird song. Like the information in those simpler sources, however, this information is carried only to those who know how to interpret it. If you give me a recipe that is written in Russian, then it will carry no information to me whatsoever; likewise, an innumerate child will not know what information is carried by a clock or a timetable. To react appropriately to a simple information source like a smell, we need to have learned to associate it with a particular food, and to react appropriately to a complex information source like a recipe, we need to have learned to associate the words that it contains with their meaning. We need, in other words, to have learned to read the language in which it is written – or, in the case of our other examples, to have learned the logic of a timetable's layout, or the horological system that a particular clock uses. The information in these sources has no meaning outside the system in which it is represented, and this means that, without a grasp of that system, no receiver can react appropriately to the source.

Of course, there are important differences between the simpler and more complex cases. When a link is innate, like the plant's chemical switch or the monkey's predator-call association, variations in the source will not be met with any corresponding alteration in the receiver's reaction, because the link is fixed. In contrast, a more sophisticated, learned representation can be altered in response to variations in the source. This difference is neatly illustrated by the Diana monkeys, which have both types of link: an innate knowledge of the association between warning calls and some classes of predator and a learned knowledge of the association between the calls and particular species of predator. As a result of its innate hardwiring, a young monkey would presumably be unable to change its reaction to the eagle call, no matter how often it experienced leopard-like danger after hearing it, whereas it is able to change its reaction to non-predatory birds as a result of experiencing danger only from eagles.

In every informational example, however, there would be no reaction at all from a receiver that encountered the source but was unable to link the information with the corresponding reaction. Jablonka and Lamb

(2005: 54) have said that, for a source to carry information, "there must first be some kind of receiver that reacts to this source and interprets it. The receiver can be an organism, a cell, or a man-made machine. Through its reaction and interpretation, the receiver's functional state is changed in a way that is related to the form and organization of the source." The authors are right to emphasize the importance of a receiver, without which there would be no reaction to the information. What their definition omits, however, is any explanation of the relationship between the information source and a particular reaction. What is it that enables the receiver to interpret and react to the source in a way that *does* correspond to the source's particular form and organization? In each of their examples, the answer lies in a preexisting link between the source and the reaction: the receiver, in each case, already "knows the meaning" (a phrase that will need to be interpreted metaphorically, of course, in the case of receivers like cells and flowers) of the incoming information.

In each case, this linking mechanism both derives information from the sensory or chemical input and controls the corresponding reaction. As Jablonka and Lamb (2005: 54) point out, the receiver's interpretation and reaction need not be intentional, but its responses do need to correspond to the form of the information. The receiver needs, in other words, to be in possession of something that produces *this* reaction when it encounters *that* information – and the simplest way to account for this link is to refer to it as a representation. More specifically, the receiver must be in possession of a *discrete* representation.

Discrete in a statistical sense means that a variable's values are consecutive rather than infinitesimally close. A continuous variable, in contrast, has a continuum of possible values: there are no gaps between members of a continuous set, and in mathematical terms its analysis therefore demands integration rather than summation (Dietrich and Markman 2003: 100). As Eric Dietrich and Arthur Markman (2003: 101) have shown, "A system cannot discriminate between two external, environmental states with one, single continuously varying representation." In order to distinguish between two points on a continuum, S1 and S2, the system needs to categorize these different inputs: it must somehow elide "the continuous infinity of intermediate states" between the two points, by forming representations that "chunk all the states in some neighborhood of S1 with S1, and all the states in some other neighbourhood of S2 with S2" (2003: 101). This means that the system is unable to discern the difference between inputs that are all in the right neighbourhood, even when they do in fact differ from one another; but "the benefit of losing information from continuous representations is the production of a set of discriminating, potentially referring, discrete representations that are combinable" (2003: 112).

In contrast, a system with one continuously varying representation is not able to discriminate between different environmental states. This is why the dial on the thermostat could turn all day, in the absence of a tappet, without any effect at all on the central-heating system: the dial's turning is a continuously varying representation of the continuously varying time of day, but the tappet is a discrete representation of a particular time of day, enabling the system to link the dial's position to the appropriate action. Similarly, an individual might have a continuously varying representation of the smell in a restaurant, the state of the weather or the sounds in his garden, but he will only identify a particular food, the likelihood of rain or a particular bird if he also has a discrete representation of a particular smell, cloud pattern or bird song. Vervet monkeys need to form a discrete representation of a particular kind of predator in order to act on it; the formation of a continuously varying representation of the other animals around them is not sufficient. A reader needs to form discrete representations of the words in a sentence if she is to understand their meaning: it is not sufficient for her to have a continuously varying representation of the marks on a page.

A discrete representation, or in more complex cases the knowledge of a representational system, enables the receiver to link variations in the source to the appropriate variations in its response. Although the central-heating system's tappets are installed and positioned as part of an inanimate, mechanical system, functionally they play the part of a discrete representation. The same is true of a photoperiodic plant's evolved, chemical switch and of a person's associatively learned link between a smell and its food source. In more complex cases, like language or notation, we need to learn not one link between a source and a reaction but an interrelated web of links between a range of sources and the appropriate reactions: a whole system of representation. It is this knowledge of a relevant representational system that enables the receiver to respond appropriately to the source, interpreting and reacting to its informational variations.

It is perhaps counterintuitive to describe a plant's photoperiodic response mechanism as representational knowledge, but some representations do play a role very much like a switch, simply linking an organism's perception of a given stimulus to behaviour that is appropriate as a response (Distin 2005: 34). Other representations are more complex, of course, and have not only these external links to perceptions and behaviour but also internal links to other representations. The evidence shows, for example, that the Diana monkey does not make a direct link between an alarm call and the appropriate evasive action: there is also an internal link to a representation of the relevant predator. In the most complex cases of all, like literacy, information is extracted from a source not by a single representation but by knowledge of a representational system. In every case, however, what matters is that a receiver should be able to link the source with its meaning.

Support for this view comes from a surprisingly varied combination of authors. From his background in biological anthropology and linguistics, Terrence Deacon (1997: 92–3) stresses that it is impossible to gain information from a representation if you do not already understand the system of which that representation is a part: a "logically complete system of relationships among the set of symbol tokens must be learned *before* the symbolic association between any one symbol token and an object can even be determined." The mathematician Keith Devlin (1999) makes a comparable distinction between the data that are found in charts or graphs and the information that a knowledgeable person can glean from those data. Their combined knowledge of narrative, mathematics, and biology persuades the authors Terry Pratchett, Ian Stewart and Jack Cohen (2003: 188) that information depends on "a protocol for turning meanings into symbols and back again": a tree's rings don't provide information to anybody who hasn't worked out the rules, and "[w]e wouldn't be able to work out our ancestry from the fossils that we have discovered unless we'd learned just what clues to look for" (Pratchett, Stewart and Cohen 2003: 109). These authors emphasize, also, that the same information source can provide different information to receivers who are looking for different clues because they have worked out different rules.

For this reason, representational systems introduce a whole new level of reliability and transgenerational persistence to the inheritance of any information or resource. The information that a receiver acquires from any source will depend on how the receiver represents it. If the source is nonrepresentational (like an artefact or an observed situation), then it has no fixed representational content, but information may still be acquired from it, depending on how it is represented by each receiver. If the source is representational (like spoken language), then a receiver would need to know the relevant representational system in order to acquire the information that is represented in it. Without knowledge of that system, the receiver might acquire some information (e.g., this script looks Cyrillic, Hebrew or Devanagari) but not the source's representational content: it is knowledge of the system which enables a receiver to acquire whatever content the source represents. This means that representational content is more persistently and reliably heritable than the information that receivers might acquire from nonrepresentational sources, because representational content is guaranteed an inheritance mechanism wherever there is a population of receivers that share the relevant representational system.

Prescriptive versus Descriptive Information

If representations form links between sources of information and the corresponding reactions or interpretation, then there is no meaningful distinction between prescriptive and descriptive information. This conclusion is significant for evolutionary theory, which has its roots in our understanding of heritable information. The claim is sometimes made that genes are unique in having been naturally selected to play a special causal role in developmental systems: a causal role that is distinctively instruction-like, in a way that is not ascribed to environmental conditions even when they are predictively important (Godfrey-Smith and Sterelny 2007). If this were the case, then we should perhaps expect to find an analogously special kind of information at the heart of cultural evolution. At the very least, we should be able to distinguish between the examples that have been discussed in this chapter, on the grounds that some of the information sources play this instruction-like causal role in the production of the receiver's reaction, whereas others do not. This section will show that genes are not causally special by virtue of being uniquely prescriptive, because all information is effectively prescriptive. Nonetheless, this conclusion does leave open the possibility of causally special factors in evolution.

On the one hand, it is true that we can detect in examples like DNA, recipes and train timetables a prescriptive relationship between the information source and the receiver's reaction to it: amino acid sequences are prescribed by information in the DNA; the nature of the apple pie is prescribed by information in the recipe; the times at which trains run are prescribed by the information in the train timetable. As a result, the organism's correct implementation of the DNA instructions will result in a particular kind of amino acid; the cook's correct implementation of the recipe's instructions will result in a particular kind of apple pie; the train operator's correct implementation of the train timetable will result in a particular pattern of train services.

A restaurant's food, on the other hand, is not prescribed by its smell. It makes no sense to say that the customer's correct implementation of the instructions in the smell will result in a particular kind of food, or that changes in the smell's instructions will produce changes in the food – because there *are* no instructions in the smell. Similarly, the time is not prescribed by a clock. A human's correct interpretation of a clock face will not result in the arrival of a particular time of day. How significant is this distinction between prescriptive and descriptive information? Although we might well have an intuition that genes play a somewhat privileged role in evolution, it is clear that an account of information, which includes recipes (because they *prescribe* ingredients and methods) but excludes clocks (because they only *describe* the time) would not be very convincing.

The conundrum is resolved by a closer examination of the nature of representations, which reveals that the way in which we intuitively distinguish between prescription and description is not precise enough for the purposes of information theory. Conversationally, we say that information is *descriptive* when it describes what a situation is already like, and *prescriptive* when it contains instructions for creating a situation that doesn't already exist. But: describes it to whom, and instructs whom? Without a receiver, there is no information at all. It is therefore more accurate to say that information is prescriptive when it contains instructions about how a receiver should *do* something and descriptive when it contains instructions about how a receiver should *interpret* something.

The position of a clock's hands, for example, or the numbers on its display, carry no information to a receiver who does not know how to interpret those horological systems. To a receiver who does have that knowledge, the clock carries information about *how he should interpret* the current time of day. His knowledge of that clock's horological system enables him to make a link between its current display and a particular interpretation of the earth's current rotational position relative to the sun. Unlike a recipe, which carries information about how to create a particular dish, the clock does not carry information about how to create a particular time of day: changing the clock display will not change time in the way that altering a recipe would alter the dish. But changing the clock display will change a receiver's interpretation of the time (which will often, incidentally, have an impact on his actions).

This distinction is illustrated particularly clearly by examples in which the same physical source can carry both types of information, depending on the receiver. A train timetable, for instance, carries information to the train operator about the times at which trains *should* run: in conventional language, it *prescribes* their schedule. It also carries information to the passenger about the times at which the trains *do* (allegedly) run: in conventional language, it *describes* their schedule – but in informational terms we can see that in this role it *prescribes* the passenger's *knowledge* of the train schedule. Does she regard a train that arrives at 9.15 a.m. as running on time or as being twenty minutes late? Her interpretation has been prescribed by the train timetable.

Information does not exist without a receiver to link input to output, and it is the nature of that link – the receiver's representation of the input – that shapes the output. In this sense, *all* information is prescriptive, even if in conventional language we should describe some of it as descriptive. But it is important to bear in mind that receivers' reactions can be prescribed by the information that they acquire from a source only if they already "know" how to discretize that information. To receive meaningful information from a source, receivers must have representational knowledge of the responses that are appropriate to any changes in the source. And as Dietrich and Markman (2003) have shown, their representations of the source must be discrete if they are to react to its changes.

Genes: First among Causal Equals?

If all of this is true for every type of information, then perhaps it is not fair, after all, to assign genes a privileged causal role in organisms' development and evolution. As developmental systems theory has emphasized (see, e.g., Jablonka and Lamb 2005; Oyama 2000), DNA is not the only resource that is transmitted through life cycles, or indeed the only factor in developmental construction. And if genetic information is not causally privileged in biology, then perhaps we should not expect to find that cultural information is causally privileged, either.

Certainly, information's dependence on the right kind of receivers can be taken as confirmation that information alone will not be causally sufficient in any sphere. But we need to remember that causal equality, in terms of the necessity of each causal factor for the outcome, should not be taken to imply explanatory symmetry, in terms of the difference that each factor makes to the outcome (Clark 1998b). In other words, even if we need to invoke a whole range of causal factors to explain *how* an outcome was produced, we may need only one of those factors to explain *why* the outcome was what it was.

An individual human's brown eyes, for example, are the developmental outcome of a range of genetic and environmental factors, all of which must be taken into account if we wish to explain how she came to have brown eyes. If, however, we wish to explain why she has brown rather than any other colour eyes, then we may need to appeal to only some of the causal factors that were necessary to produce that developmental outcome. Clark (1998b: 161) advocates that in order to establish whether any of these factors is causally privileged, we need to investigate "the extent to which difference in respect of that outcome within a baseline population and ecological setting may be traced to difference in the privileged item." In this case, we may well conclude that genes have played a causally privileged role, in the sense that variations in eye colour within a typical human population can be traced to variations in the relevant genes. This is not to say that genes are the only factor, or even the most necessary or significant factor, in the developmental construction of human eyes. It is rather to claim that there is no contradiction between maintaining both that there are many necessary causal factors in the developmental construction of human eyes, and that genetic variations should be causally privileged in explanations of variations among individuals' eye colours.

Nor is it to assume that it will always be genes that are causally privileged in our explanations. Indeed, we should be warned off this assumption by the significance that this chapter has afforded to the receiver in its characterization of information: we might well expect that variations in developmental outcomes will sometimes result from variations in the receivers of genetic information as well as from variations in its content. Information in any context cannot exist in isolation, but always depends on a receiver that can discretely represent and consequently interpret and respond to it. For this reason, the receiver's responses will be determined by a range of causal factors, including the nature of the information and the nature of its representation(s), as well as the environment or context. Whether one or more of these factors should be causally privileged in our explanations of each receiver's response will be open to question in each individual case. How information is transmitted between receivers, what enables the receivers to respond to it, and what role these transmission mechanisms play in evolution are the subjects of the next chapter.

3

How Is Information Inherited?

Evolution in any sphere results from the persistent heredity of variations. What has emerged from the previous chapter is that each generation can detect and react to inherited variations only if it can discretely represent them – and any variation that a receiver discretely represents is information. It turns out that information is indeed "a difference which makes a difference" (Bateson 1972: 453).

The inheritance of information is therefore dependent on the representational capacities of each generation of receivers. We have seen that it is the receiver's *discrete representational knowledge*, whether in the form of a single representation or of an overarching representational system, which enables it to react appropriately to a source of information. This means that the interpretation and effects of information will depend not only on the content of the information itself but also on variation among its receivers. If a receiver cannot form discrete representations of variations in the information that it receives, then it will be able neither to react to those variations nor to transmit them to another receiver.

If heredity depends on successive generations of receivers with the appropriate representational knowledge, then the mechanisms of heredity need to ensure that each generation receives not only information but also its means of interpretation and implementation. Jablonka and Lamb (2005) have proposed a four-dimensional account of biological evolution, in which not only genetic but also epigenetic, behavioural and symbolic inheritance mechanisms play their part. What we need to ask about any purported inheritance mechanism is whether, and how, it enables the next generation to respond to the available information. This account of heritable information can go some way towards resolving the debate between what are sometimes called standard and extended models of evolution (e.g., Odling-Smee, Laland and Feldman 2003), and it enables cultural evolution to be brought under the same explanatory umbrella as biological evolution.

Genetic Inheritance Systems

The paradigmatic example of an inheritance system is, of course, genetic[1] – but although nobody would deny the importance of genes for biological evolution, there is some disagreement among biologists about just how privileged a role they play. We can usefully talk about DNA as self-replicating, for example, but it is important to remember that genes cannot work in isolation: genes are not, strictly speaking, replicators but rather replicas. Terrence Deacon (1999: 4) gives a characteristically clear summary of the situation:

> Genetic replicators are just strings of DNA sequence information that happen to get copied and passed on to the future successfully. Genes are not active automatons, they are not "agents" of evolution, just structures, passive information carriers that can become incorporated into biochemical reactions, or not, depending on the context. Referring to genes *as though* they were active agents promoting their own copying process is a shorthand for a more precise systemic account of how a gene's effect on its molecular, cellular, organismic, social, and ecological context influences the probability of the production of further copies. Taken out of this almost incomprehensibly rich context – placed in a test tube, for example – DNA molecules just sit there in gooey clumps, uselessly tangled up like vast submicroscopic wads of string.

In this sense, developmental processes are as important for an organism as the DNA whose instructions they follow. When you were conceived, you inherited not only nuclear DNA from both parents but also the cytoplasm from your mother's ovum. Without the cellular mechanisms of the maternal cytoplasm, the nuclear DNA would have no effects.

DNA carries information, in a sequence of four types of nitrogenous nucleotide bases, about a sequence of amino acids in proteins. These nucleotide bases are grouped in threes, called codons, each of which specifies an amino acid. There are some interspecies variations in the standard genetic code: the same codon can mean different things in different organisms. The coding mechanisms, however, are the same for all species: the information in DNA is rewritten in another script, RNA, which carries information from the nucleus to the cytoplasm, where

---

[1] I am grateful to Professor L. S. Shashidhara (personal communication) for greatly increasing my understanding of the biological processes described in this section.

proteins are generated according to the information provided by DNA in the form of RNA. The machinery that assembles proteins, by translating the genetic information codon by codon into sequences of amino acids, is provided by complexes of RNA and proteins, called ribosomes. Although all of the cells in an organism contain the same DNA, each expresses a different set of genes, or makes different sets of proteins, which confer each cell's individual identity. This differential gene regulation is facilitated by another set of proteins, known as transcription factors. These transcription factors "tell" the transcription machinery which part of the organism's genetic information to read.

In effect, we can see the cell's transcription and translation machinery as its "knowledge" of the genetic language, without which there would be no appropriate reaction to the codons. Every time a cell divides, the information that it contains in DNA is faithfully reproduced in the same language and format, but it could not be expressed if the cellular machinery for its transcription and translation were not also reproduced. Like any other representational system, DNA's expression depends on the existence of a receiver for the information that it carries: a receiver that knows the system. A daughter cell is able to express the base sequences that it received from its mother cell because it knows what those base sequences mean: it inherits not only the base sequences themselves but also the cellular decoding mechanisms without which the DNA would be useless. The mechanisms of cellular division ensure the persistent heredity of both patterns of DNA sequences and the cellular machinery for their interpretation and expression.

Given the appropriate environmental context, the genetic information that is transmitted to the next generation will be expressed, by the cellular machinery, in the associated phenotypic effects. These effects can influence the differential survival and reproduction of the genes with which they are associated, via the processes of natural and sexual selection. Genetic variations can also be neutral with respect to fitness in a given context, and in this case gene frequencies might change as a result of chance events rather than selection. Over multiple generations, these changes in organisms' genetic and/or phenotypic patterns give rise to evolution in taxonomic patterns. The mechanisms of genetic inheritance,[2] which ensure the persistent heredity of variations in genetic

---

[2] For all uses of this phrase where I do not specify otherwise, it should be read in the broadest sense to include the whole system of cellular reproduction, not just the replication of nuclear DNA.

information, as well as of their interpretative machinery, thus provide an effective process for biological evolution.

Epigenetic Inheritance Systems

Since the expression of genetic information depends on its transcription and translation by the cellular machinery, it is perhaps unsurprising to find that variations in expressed traits can be the result either of genetic alterations or of what are known as epigenetic changes in the ways in which the cellular decoding mechanisms interpret genetic information. If, for example, DNA is chemically altered by the addition of methyl groups that can be sensed by transcription factors, or if there are changes in the proteins around which chromosomal DNA is wrapped, then different meanings can be produced from the same DNA sequence. Normally, parental epigenetic marks of this kind are "wiped out and reset" at fertilization, but when there is disruption to this process of epigenetic reprogramming, so that the parental epigenetic marks are not completely erased, the result can be "transgenerational epigenetic inheritance" (Xing et al. 2007): the DNA transmitted is identical, but variations in its expression are inherited across cellular generations.

For instance, it has been shown (Yehuda et al. 2000; Yehuda, Halligan and Bierer 2002) that the adult children of Holocaust survivors, and more recently (Yehuda et al. 2005) that the infant children of September 11 survivors, have lower cortisol levels, which in both cases alters their stress response. This is especially the case when their mothers were traumatised during the third trimester of their pregnancy (Yehuda et al. 2005), but the same effects are seen in children who were born years after the Holocaust to mothers, but not to fathers, of survivors (Yehuda et al. 2008) – as would be expected if hormonal markers were created in the ova of traumatised women. Recent research also indicates that the prepubescent environment can affect boys' sperm, with consequent effects on the next generation. One study found, for example, that "early paternal smoking is associated with greater body mass index (BMI) at 9 years in sons, but not daughters"; and that a paternal grandfather's childhood food supply has an impact on grandsons' but not on granddaughters' longevity, whereas a paternal grandmother's food supply is associated only with granddaughters' longevity (Pembrey et al. 2006).

A raft of recent research indicates the previously unappreciated importance of epigenetic mechanisms for the regulation of gene expression in response to environmental signals, and provides evidence of the trans-generational transmission of epimutations (Anway and Skinner 2006; Cavalli and Paro 1998; Hitchins et al. 2007; Xing et al. 2007). "The significance of this phenomenon," however, "and the mechanism by which it occurs, remains obscure" (Xing et al. 2007). Although parental epigenetic signals to offspring about the environment are by definition transgenerational, to date there is no consensus about whether epigenetic changes persist beyond one or two generations. There is some support for the persistence of transgenerational epigenetic effects, at least to the fifth generation (Anway and Skinner 2006; Cavalli and Paro 1998), but other studies have found that epigenetic effects are diluted after a couple of generations (Xing et al. 2007). It seems likely that at least in some cases they are available only in the short-term, and from an evolutionary point of view this makes perfect sense: detailed environmental information is likely to be useful for only a limited number of generations. The inherently temporary nature of epigenetic changes, in these cases, isolates them from evolution in a way that introduces a selectively useful level of flexibility to an organism's development. Genetically in-built flexibility of this sort is a standard evolutionary strategy, and the length of the gene-phenotype leash varies according to the rate of environmental change.

The situation is clearly complex, and as yet poorly understood. At this stage, from a nonspecialist's perspective, it is safest for me to say that, although currently it is uncertain whether epigenetic mechanisms can support sufficiently persistent heredity to influence biological evolution directly, they do at the very least provide genes with the means of responding selectively, in a way that persists for a few generations, to some environmental changes. Even if they persist for only a single generation, it will mean that "germline epimutations must now be taken into account when looking at disease, along with DNA sequence mutations and environmental factors" (McDonald 2007). The unresolved question, then, is not whether epigenetic mechanisms are important or interesting, but how epigenetic inheritance influences the course of evolution. Can epigenetic inheritance mechanisms ever ensure that a variant interpretation of a genetic sequence is inherited over sufficient generations to make it available to selection? Or does their significance lie, rather, in the medium-term developmental flexibility that they give the genome?

Either way, we can see that the mechanisms of cellular reproduction produce a daughter generation that receives both genetic information

and representational knowledge of the code in which that genetic information is written. Phenotypic variations might be the result of variations
in either the genetic information or the ways in which it is interpreted, or
indeed the environment. Genetic inheritance systems ensure that variations in DNA patterns are transmitted to the next generation. Epigenetic
inheritance systems usually ensure that variations in epigenetic patterns
are reset when organisms reproduce, but the possibility and extent of
transgenerational epigenetic inheritance is currently up for debate. Both
genetic and epigenetic processes, however, are part of the same cellular
system: a system in which genetic information is replicated and expressed
by cellular machinery that is also received by each new generation.

In the case of DNA and its cellular machinery, then, the same mechanism is responsible for the heritability of both information and its means
of interpretation and expression. When, however, we turn to developmental and ecological systems of inheritance, via which resources other
than DNA are transmitted through life cycles, we find that these resources
depend for their inheritance on one mechanism, whereas a different
inheritance mechanism ensures that the next generation can exploit
them.

Ecological Inheritance Systems

Variations in the phenotypic expression of genetic information can be the
product of environmental as well as of genetic or epigenetic variations.
But just as the environment affects organisms' growth and survival, so
organisms themselves can modify the environment, in a process known
as niche construction. Ecological inheritance mechanisms are said to be
at work whenever the physical consequences of one generation's niche
construction persist in the environment, so that the next generation
receives a legacy of "modified natural selection pressures" (Odling-Smee
et al. 2003: 13), and subsequent generations receive a similar resource
because of the activities of the current generation, and so on. A beaver,
for example, inherits an environment that has been created by the dam
that past generations built; but it also, by working on the dam, maintains
that environment for future generations.

It can sometimes seem that there is a theoretical chasm between an
extended evolutionary theory, which takes into account ecological as well
as genetic inheritance systems, and "standard" evolutionary theory, which
is supposed to exclude any inheritance mechanism except the genetic
system and any directing process except natural selection (Odling-Smee

et al. 2003). The chasm can be bridged, however, by the model of heritable information that has been developed here. Nobody denies that organisms modify their environments, or that environmental modifications sometimes persist for long enough to be received by the next generation. The interesting question, according to my theory of inheritance, is what enables the next generation to act as a receiver of those modifications.

It is clear that beavers, for instance, are able to use and maintain their dams because of their genetic endowment, and in this sense, as Richard Dawkins (1982) has pointed out, beavers' dams, spiders' webs and similar natural artefacts can be seen as an extended part of these creatures' phenotypes. The difference between a beaver's fur and its dam, however, is that its fur is the expression of its own genes, whereas the dam is an expression of its parents' genes. In this sense, although it inherits variants in both fur and dam from its parents, the inheritance mechanisms are different in each case. The beaver's individual DNA patterns, and the cellular machinery for their interpretation and expression, were inherited via the mechanisms of sexual reproduction. This tells us that the transgenerational persistence of variants in beaver fur is the product of genetic inheritance mechanisms. The beaver's environment, on the other hand, was altered by its parents as an expression of their genes, and this niche-constructing *behaviour* is the mechanism that ensures the transgenerational persistence of variants in beaver dams.

Thus, ecological inheritance mechanisms like the beavers' damming behaviours can indeed act as a mechanism of ecological inheritance: environmental modifications are transmitted from one generation to the next by the niche-constructing behaviours of each generation. If we ask what resources each new generation receives from its parents, then our answer must include both genetic resources, which are received via genetic inheritance mechanisms, and environmental resources, which are received via ecological inheritance mechanisms. Does this mean, as some authors have claimed, that "standard" evolutionary theory is no longer good enough, and that we need to extend not only our understanding of the phenotype (Dawkins 1982) but also our understanding of evolutionary theory itself (Odling-Smee et al. 2003)? To answer this question, we need to know whether evolution can result from ecological inheritance mechanisms in the way that everybody agrees that it can from genetic inheritance mechanisms.

An important distinction is revealed, between mechanisms of inheritance and mechanisms of evolution, when we ask *how* each generation of niche-constructing organisms is able to act as a receiver of the environmental modifications that its parents have made. Like any other resource, the environment can be interpreted and exploited in different ways, depending on the nature of the receiver. Environmental changes will not be informative to an organism that is not "prepared" (Dickins 2005: 81) to respond to the changes, and in this circumstance the organism will not be able to act as a transmitter of those changes. To pass a resource like a dam through generations of beavers, each generation has to transmit not only the dam itself but also the capacities to benefit from that resource, to maintain it and pass it on to the next generation. Otherwise, the transmission would be only one generation deep. Without the biological inheritance mechanisms that enable the next generation to respond appropriately to the environment that it inherits, therefore, a beaver's damming behaviour would be effective as an ecological inheritance mechanism over only a single biological generation. Evolution depends on inheritance mechanisms which ensure that variations are passed on through multiple generations.

Thus, although there is general agreement that biological evolution can result if the environment is modified in such a way and over a sufficiently long period of time that new adaptive challenges emerge for the niche-constructing organisms, there is less agreement about whether this counts as a new evolutionary mechanism. On some views, what's happening here is conceptually no different from the familiar process of natural selection (Dickins 2005: 81–2). The problem, as has been noted, is that even quite drastic environmental change will not result in evolutionary change in the local organisms, unless those organisms are able to detect and react to what has happened. The environmental changes are explicable by ecological inheritance mechanisms, but we need to appeal to genetic inheritance mechanisms in order to explain how the organisms are able to make an evolutionary response to the new adaptive problems that have emerged.

It is only once we ask *how* one generation is able to act as a receiver of resources from another that the tangle of evolutionary mechanisms and driving forces begins to unravel. The answers reveal that niche-constructing behaviours can act as mechanisms of ecological inheritance, transmitting environmental changes from one generation of niche-constructing organisms to the next; and they can also act as a mechanism of long-term *environmental* change. What ecological inheritance mechanisms cannot do is act as a mechanism of *biological* evolution, unless they are combined with the usual processes of genetic inheritance.

Developmental Inheritance Systems

A similar consideration applies to what Jablonka and Lamb (2006: 12) describe as "body-to-body" inheritance: a process in which "offspring receive materials from their parents, which lead them to reconstruct the conditions that caused the parents to produce and transfer the material to them, and thus pass on the same phenotype to their own descendants." For instance, the authors cite experiments with European rabbits, which show that "[r]abbit pups raised by mothers fed different diets during pregnancy and lactation show a clear preference for the diet of their mother at weaning" (Bilkó, Altbäcker and Hudson 1994: 907). The experimenters investigated "the relative importance of 1) fecal pellets deposited by the mother in the nest, 2) prenatal experience in utero, and 3) contact with the mother during nursing in determining pups' later food preference. The three means of transmission were found to be equally effective."

The first question that we need to ask, once again, is what enables pups to make sense of this information – and once more, we find that the answer lies in their genes. If the pups did not have an innate ability to link uterine and neonatal nutrition with future dietary preference, then the maternal products would not carry any information to them. The pups' suckling and pellet-eating behaviour is innate, and so is their ability to interpret and react to the information that their mother's milk and pellets contain. It is this innate link – this representation, if you like, of the dietary information in the maternal products – that enables the pups to receive and react to the information. They are innately disposed to suckle and to eat their mother's faecal pellets, and these behaviours expose them to the dietary information that is available in the maternal products.

Although they can make sense of that information because of an ability that they have received via the genetic inheritance mechanism, however, they receive the developmental information itself via a developmental inheritance mechanism: their suckling and pellet-eating behaviours. If we ask what resources each new generation of rabbits receives from its parents, then our answer must include both genetic resources, which are received via genetic inheritance mechanisms, and dietary information, which is received via developmental inheritance mechanisms.

To what extent can this type of inheritance provide a mechanism for evolution? It seems clear that the rabbits' developmental inheritance mechanisms have a more limited capacity to carry cumulative changes

through multiple generations than have the ecological inheritance mechanisms that were considered in the previous section. Variations in one
generation's diet, at a particular time in its life, may well not be transmitted to the next generation. This is amply demonstrated by the very
experiments that showed the relationship between maternal diet and
pups' preferences at weaning: the pups' preferences were influenced
equally by their mothers' diets in utero, during nursing and when they
were old enough to eat her faecal pellets, ensuring that they "can acquire
a preference for a variety of foods eaten by their mother at different
times" (Bilkó, Altbäcker and Hudson 1994: 907). Such genetically in-
built flexibility ensures that a novel variation in the maternal diet can be
passed on to the next generation, which in turn might pass on its own
novelty to the subsequent generation. There is no guarantee, for example, of a match between the dietary preferences that a pup learns from
its mother and the availability of those foods during its own subsequent
pregnancy and lactation.

This is not to say that developmental inheritance systems can never
result in the long-term evolution of developmental resources. On their
own, developmental inheritance mechanisms like the rabbits' suckling or
pellet-eating behaviours can transmit information to organisms that are
genetically prepared to receive it, and they may, in some circumstances,
be able to act as a mechanism of long-term dietary change. Biological
evolution could result from this process if the rabbits so changed their
diets that new adaptive challenges emerged – but only if the rabbits
were genetically prepared to detect and react to what had happened.
Developmental inheritance mechanisms cannot act as a mechanism of
biological evolution unless they are combined with the usual processes
of genetic inheritance.

Behavioural Inheritance Systems

When the rabbit pups acquire dietary information from their mother's
bodily products, they are engaged in a process of individual learning.
Information can also be transmitted between receivers by socially mediated learning mechanisms, and this more flexible kind of learning is especially useful when it comes to "aspects of the environment that change
too rapidly for the main genetic program to track, such as change that
occurs within generations" (Mesoudi 2008: 247). Individuals can learn
from conspecifics both what to do and how to do it. A particular focus

of Jablonka and Lamb's work in this area is on the long-term effects of socially mediated learning across multiple generations, as seen in animal traditions: "behaviour patterns that are characteristic of an animal group and are transmitted from one generation to the next through socially-mediated learning" (2006:14–5).

It seems clear that social learning can provide the inheritance mechanism for stable complexes of animal behaviours (Jablonka and Lamb 2006: 15): so long as group members are innately disposed to learn from one another as youngsters, the things that they learn will automatically be passed on to the next generation, along with any embellishments or changes that are made either during the copying process or later, during an individual's lifetime. There is no doubt that these organisms can receive information via behavioural as well as genetic inheritance systems, and the ethological evidence indicates that these behavioural inheritance systems can result in such long-term persistence in behavioural patterns that we might feel justified in referring to the outcome as behavioural evolution.

Can they also result in evolution in the social-learning organisms themselves? There are several reasons why the behavioural inheritance system is unlikely to provide an additional mechanism for biological evolution. First, there is the familiar problem that behavioural inheritance mechanisms cannot provide each new generation of organisms with the capacity to receive the information that they transmit. For this, organisms need an innate capacity to engage in social learning, as a result of which they are genetically prepared to receive intraspecies information about behavioural patterns – and this capacity is genetically inherited. Relatedly, although it is not implausible that the persistence of these behavioural patterns should give rise to new adaptive problems, it remains the case that the organisms' capacity to detect and react to these changes is dependent on the genetic inheritance system. Finally, unlike ecological or developmental inheritance systems, which are largely restricted to routes through genetic descendants, behavioural inheritance systems enable group members to learn traditions from a variety of conspecifics: the fact that a fitness-enhancing tradition is transmitted to one individual, therefore, does not necessarily increase the fitness of that individual's offspring to any greater degree than it increases the fitness of any other individual's offspring. Evolution is founded on the inheritance by offspring of parental variations, but the behavioural inheritance system is not restricted to the parent-offspring route. For these reasons, the behavioural inheritance system cannot provide a separate mechanism for biological evolution, even when it can provide a mechanism for changes in group behaviours.

Genetically Mediated Inheritance

Various mechanisms have been proposed for the transmission of information and resources through life cycles, and it is apparent from even such a brief review as this chapter has given that the genetic system is not the only inheritance mechanism. Genetic information and its epigenetic translation systems are inherited by the mechanisms of cellular reproduction. Developmental and environmental resources can be inherited by the mechanisms of genetically controlled behaviour, including niche construction and both individual and social learning. Each of these systems carries intergenerational information at a different speed and with a different level of persistence.

Genetically transmitted information has the advantage of persistent heredity through countless generations – and the disadvantage of persistent heredity through countless generations, for if the environment changes too quickly, then genetic information may not be able to adapt quickly enough to keep up with the novel demands. Epigenetic mechanisms can produce variations in response to these medium-term changes, but it is unclear whether they can persist beyond a few generations. Ecological, developmental and behavioural inheritance mechanisms can transmit environmental, developmental and behavioural variations from just one generation to the next, with the advantage that they can produce variations in response to changes that happen within a single generation – which might then be obliterated by changes in the next generation's environment.

None of the last three inheritance mechanisms can serve as a mechanism of biological evolution, because heritable variation depends not only on the effective transmission of the variants themselves but also on the effective transmission of their means of interpretation. A receiver must be able to detect and react to variations in a source in order to acquire and transmit its information: an organism must, in other words, be prepared to make use of the available input (Dickins and Dickins 2008) – and it is the genetic inheritance system that puts this preparation in place.

What matters for evolution is that variations persist through a sufficiently large number of generations to be selected, and that they be capable of causing design change in that which is inherited. Ecological, developmental and behavioural inheritance mechanisms cannot ensure that ecological, developmental or behavioural changes persist across sufficient biological generations to affect the course of biological evolution. What they can (sometimes) do is ensure that ecological, developmental or behavioural changes persist across sufficient *ecological, developmental* or *behavioural* generations to affect the course of ecological, developmental or behavioural evolution. Rather than providing additional dimensions of biological evolution, then, these inheritance mechanisms may sometimes provide additional tokens of the evolutionary type.

Linguistic Inheritance Systems

There is no doubt that human language use provides an inheritance mechanism that enables information to be transmitted from one human generation to the next. Humans who use languages provide a resource for the next generation, just as much as beavers that provide dams or rabbits that provide bodily products for their offspring. The question that remains up for debate is whether cultural inheritance mechanisms provide another dimension of biological evolution (Jablonka and Lamb 2005): another system by which information or resources can be transmitted from one biological generation to the next, ultimately facilitating evolutionary change in human biology. Cultural evolutionary theory agrees with Jablonka and Lamb that the mechanisms of cultural inheritance are systems of symbolic representation, which are uniquely human and include not only language but also the symbol systems of mathematics, music and so on. It claims, however, that cultural inheritance provides a mechanism for cultural rather than biological evolution.

To discover whether cultural inheritance mechanisms result in biological or cultural evolution, we need to ask how these inheritance mechanisms work. We have seen that an inheritance mechanism that transmits only information or resources cannot, on its own, act as a mechanism of evolution: the transgenerational persistence of the variations that it transmits will depend on each generation's ability to detect and react to those variations. So we need to understand both how symbolic systems transmit information between human receivers and how humans are able to make sense of the symbolic information that they receive.

Information, as we have seen, can only be transmitted to a receiver that knows how to make sense of it: the receiver must be able to link variations in the source to the appropriate variations in its response. In the case of a complex source like language, this means that a receiver must understand a whole system of representational links. As Dietrich and Markman (2003) have shown, it is not possible for a receiver to discriminate among continuously varying states of an environmental source without discretizing that source: human cognition must discretize a linguistic source in order to process it effectively. A continuously varying input, like a sound wave, might be represented continuously at an early stage of auditory processing, but higher-level processing systems must extract information in chunks from this continuously varying representation, in order to produce discrete representations that discriminate between parts of the continuous input (Dietrich and Markman 2003: 112). Humans cannot receive cultural information until they have learned to discretize the language in which it is represented.

Even the most hard-line Chomskian would not claim that humans are born with a genetically supplied knowledge of a particular natural language, sufficient to enable them to respond appropriately to the linguistic input that they hear from birth. Rather, humans have a genetically supplied ability to form such representational links through learning: languages are transmitted between humans by a process that combines genetic with linguistic inheritance mechanisms. When the current generation learns the local language, it re-creates that linguistic resource for the next generation, but it relies on genetic inheritance mechanisms to ensure that the next generation is able to exploit and maintain that resource. This is a very similar process to the physical niche construction of a species like the beaver: linguistic inheritance mechanisms, like ecological inheritance mechanisms, ensure that certain resources are transmitted from one generation to the next, but they are dependent on genetic inheritance mechanisms to ensure that the next generation is capable of acting as a receiver. Just as ecological inheritance mechanisms can give rise to long-term changes in the physical environment, so linguistic inheritance mechanisms can give rise to evolution in languages. Given the availability of niche-constructing beavers to act as receivers of the ancestral environment, niche-constructing behaviour can act as a mechanism for environmental change. And given the availability of language-ready humans to act as receivers of the previous generation's language structures, language-learning behaviour can act as a mechanism for language evolution.

As with ecological inheritance mechanisms, however, linguistic inheritance mechanisms cannot directly give rise to biological evolution. Although human language use has doubtless transformed the adaptive problems that humans face, we should still need to appeal to genetic inheritance mechanisms to explain any evolutionary response that humans might make to the adaptive landscape. Moreover, although it might seem intuitively obvious that languages are transmitted primarily from parents to children, in reality children will hear their native language spoken by most of the people they meet, and a child who is adopted at birth will not engage in any postnatal language learning from her biological parents. Languages are transmitted across linguistic rather than biological generations and are not restricted to genetic routes; consequently, their evolution is disconnected from, and far outpaces human biological change.

Of course this argument leaves open all sorts of questions about human languages and their evolution. To what extent is language acquisition innate, providing biological constraints on the directions of language evolution? Why are humans unique in acting as instinctive receivers of language structures, and what are the evolutionary roots of our language capacity? These are questions for subsequent chapters (see especially Chapters 4 and 5), but for now I have sketched my claims with such a broad brush that their truth is unaffected by the details of particular theories of language. These claims are that language relies for its transmission on receivers who are able to acquire its structures from hearing it spoken – humans who are in some way genetically prepared to be language ready; that humans' instinctive linguistic behaviours provide an inheritance mechanism for human languages across generations of language users (which may or may not coincide with the generations of humans themselves); and that linguistic inheritance provides a mechanism for linguistic, rather than biological, evolution.

Cultural Inheritance

There is a crucial distinction between the languages that humans learn and the cultural information that those languages enable them to acquire. This distinction is brought into focus when we ask how humans are able to receive the resource of previous generations' languages, and how they are able to receive the resource of the information that those languages transmit.

We have seen that human language learning is a nongenetic inheritance mechanism just as much as beavers' dam building is: in both cases, an organism's behaviour is the inheritance mechanism for variations in that which is inherited. In both cases, however, it is the genetic inheritance mechanism which ensures that the organism can detect and respond to those variations in the first place. So although a nongenetic mechanism is responsible for the inheritance of a particular resource, the genetic mechanisms are responsible for the inheritance of the organism's ability to receive the resource.

But if humans are *genetically* prepared to learn language, then in receiving that resource, they become *culturally* prepared to acquire the information that language transmits. By enabling us to discretize their semantic content, languages effectively turn us into receivers of the information that they transmit. Once biologically language-ready humans have acquired the local language, they become developmentally culture-ready humans. What sets cultural inheritance systems apart from the other nongenetic inheritance mechanisms that we have considered is that *nongenetic* mechanisms are responsible for the whole of this inheritance process. One nongenetic mechanism (the developmental process of language acquisition) turns us into receivers of the cultural information that is transmitted by a second nongenetic mechanism (language use). In this way, human languages ensure the persistent heredity of any information that they carry, because they bring to human receivers both that information itself and the means of its interpretation and transmission. Once there is a ready supply of language-using humans to act as receivers of cultural information, their language use can provide a self-sufficient mechanism for cultural evolution – just as cellular mechanisms do for biological evolution. Cultural inheritance provides a mechanism for cultural, not biological, evolution – a conclusion that subsequent chapters will develop and defend in far more detail than has been sketched in this section.

Does Evolution Really Need Discrete Representation?

By this point, some readers may be wondering why this chapter has focused so sharply on the need for receivers to form discrete representations of the information that they inherit from a source. Although nobody denies that evolution *can* result from the differential inheritance of discrete representations, Joseph Henrich, Robert Boyd and Peter Richerson (e.g., 2008) have famously maintained that cumulative, adaptive evolution will result from any process that replicates a population's average characteristics, and that the replication of discrete, combinatorial representations of those characteristics is not the only way to achieve this. They contend that, so long as transmission is biased towards certain

representations – and the stronger the bias, the better – cumulative evolution can arise even from the inaccurate transmission of nondiscrete
representations. "Any process of cultural transmission that leads to accurate replication of the average characteristics of the *population* will work"
(Henrich et al. 2008: 11).

Recall that *discrete* in a statistical sense means that there are gaps,
however small, between a variable's values. It is certainly not the case
that cultural variation is always discrete in this sense. To emphasise this
point, Henrich and Boyd (2002: 91) consider a group of Illinois residents with varying perceptions of what counts as a fair sharecropping
contract: although most of them believe either that landowners and farmers should split the crop yield equally, or that landowners should receive
twice as much as the farmers, it is possible for them to perceive as fair
a contract in which the landowners receive any other share – like 32.76
percent – that we can imagine. There is a continuous spectrum of variation, say Henrich and Boyd, in the Illinois residents' representations of
a fair sharecropping contract, and this is one of the situations that the
authors model, mathematically, to show how continuous representations
can give rise, under certain conditions, to discrete-replicator dynamics.
Adaptive cultural evolution, in their view, is not dependent on the actual
existence of discretely replicated units, because human cognition can
produce an approximation of discrete-replicator dynamics from a continuous representational input.

The problem with this example is that it does not, despite the authors' assumption, actually involve a continuous distribution of variation.
Although there is a continuum of variation in the possible shares that
can be used in a contract, the variation in the croppers' *representations*
of those shares is discrete. This conclusion follows from the following
considerations.

Henrich and Boyd have identified a continuously varying quantity: the
number of shares that an Illinois farmer should receive for every share
that the landowner receives. There is, in theory, no reason why this number should not be any real number, and because real numbers form a
continuous set, the authors are correct to assume that this is a continuously varying quantity. The problems, for the assumption of continuity
that is needed for their mathematical modelling, arise once the farmers
begin to *represent* this quantity. As Dietrich and Markman (2003) have
shown, for the farmers to choose from a continuously varying range of
possible shares, their cognitive mechanisms must first produce, from this
continuous range, a discretized population of options. The way in which

they form their discrete representations will, of course, have an impact on the constitution of that population. If they represent numbers to only one decimal point, for example, then they will have no way of representing any number of shares between 2.2 and 2.3: the possibility of a 2.25 share, which would have been available if they had represented numbers to two decimal points, or of a $2\frac{1}{3}$ share, which would have been available if they had represented numbers as fractions, has been lost. The compensation for losing intermediate information of this kind, however, is the production of a set of combinable, discriminating representations, without which the sharecroppers cannot even begin to draw up a contract: cognition requires discrete representations.

Because each farmer has a discrete set of representations of what might constitute a fair share, and there is a finite set of farmers, it follows that their representations of a fair contract are not, in fact, continuous, even though there is a continuous distribution of possible shares. Henrich and Boyd have not shown that cultural evolution can happen in the face of continuous cultural representations, because their example does not involve continuous representations. It involves discrete representations of a continuous variable.

What Henrich and Boyd have demonstrated, however, is how human cognitive mechanisms select among available representations, ensuring the accurate transmission of those representations that it does favour and disadvantaging the alternatives. In this way, somewhat ironically, they have shown how human cognition supports the differential transmission of discretely represented cultural information.

There are two steps to this process. First, our cognitive systems need to discriminate among the continuously varying states of the environment, and for this reason, as Dietrich and Markman have described, they rely on discrete representations. Dietrich and Markman's persuasive arguments are further supported by evidence from cognitive neuropsychology that our brains store learned information in a separable hierarchy rather than blending it irretrievably. For instance, brain damage in humans can result in category-specific naming deficits (Crosson et al. 1997), and there are neurons in the human median temporal lobe whose behaviour appears to indicate that they encode for particular abstract representations: these neurons respond selectively to specific individuals or objects, even when presented with strikingly different views, or even the printed names, of those objects or individuals (Quiroga et al. 2005). Moreover, information must be discretely represented if it is to be combined with other information: "when continuous representations are combined, all that results is another continuous representation where all the original information is lost due to blurring" (Dietrich and Markman 2003: 107). Only discrete representations can be combined in such a way that the original representations are recoverable.

Languages enable us to discretize information in just this way. They are what Steven Pinker (1994: 84) has described as discrete combinatorial systems: systems in which a "finite number of elements (in this case, words) are sampled, combined, and permuted to create larger structures (in this case, sentences) with properties that are quite distinct from those of their elements." Pinker (1994: 85) contrasts such systems with what he calls "blending systems," in which "the properties of the elements are lost in the average or mixture." Indeed, it is well established that one of the most important cognitive functions of languages is the way in which they enable us to chunk their semantic content into a discrete series of units. George Miller, whose 1956 paper on this subject has been described as one of the most influential papers published in *Psychological Review* since its inception in 1894 (Cowan 2001: 87), argued that natural language recodes information to stretch the informational bottleneck of short-term memory. Recoding is a process in which "input is given in a code that contains many chunks with few bits per chunk. The operator recodes the input into another code that contains fewer chunks with more bits per chunk. There are many ways to do this recoding, but probably the simplest is to group the input events, apply a new name to the group, and then remember the new name rather than the original input events" (Miller 1956: 93). Miller (1956: 95) suggested that "the most customary kind of recoding that we do all the time is to translate into a verbal code."

So the first step that our cognitive processes take towards the differential transmission of discretely represented cultural information is to discretize the continuum of environmental states, using a combination of perceptual and linguistic processes. Henrich and Boyd's mathematical analysis shows how our cognitive mechanisms next produce, from this discretized population of mental representations, a reduced set of viable options. A recent paper, cowritten with Peter Richerson, brings this analysis up to date (Henrich et al. 2008).

Their models assume that, in a given cultural population, "every mental representation is equally common initially" – or, to put this more accurately, that there is a continuous range of shares, all of which are equally available to representation by the croppers. Once this range has been discretized, what happens during the transmission of information is that some shares are easier than others for human brains to comprehend, and these act as "cognitive attractors" (Henrich et al. 2008: 8). Croppers are much more likely to perceive a fair contract as one that involves whole numbers, for example, than they are to go for one of the many fractions in between. Such cognitive attractors act to transform people's perception of the alternatives, so that even if someone is actually using a contract that falls between the whole numbers, he will often be perceived as using one or another of the focal (whole-number) contracts (Henrich and Boyd 2002: 93). Henrich and colleagues (2008: 9) demonstrate mathematically that if the attractors are strong then very rapidly they will cause almost everyone to have a representation that is close to one of them.

In other words, even if there is a continuous spectrum of variation in a range of cultural options, Dietrich and Markman have shown that human cognition must discretize the range to represent it, and Henrich and colleagues (2008) have shown that strong attractors will quickly generate a situation in which most people's representations are near an attractor. Intermediate representations, on this view, are not persistently heritable, because they are transformed, during the cognitive processes of information acquisition, storage and retrieval, into one of the cognitive attractors.

To move out of mathematical models into the observed cultural world around us, we know that negotiations about fair shares do centre, in real life, on numbers that people can understand without too much effort. "Young and Burke (1998) have shown that an overwhelming majority of the share cropping contracts in Illinois are one of two types: In some communities the farmers and the landowners split the crop yield, while in other communities the landowners receive two shares for every one received by the farmer. Interestingly, despite wide variation in land quality within communities, there is little variation in share cropping contracts within communities" (Henrich and Boyd 2002: 93). There is no doubt that some cultural representations attract more attention than others; some are easier to learn than others; some are easier to remember than others. The structure of human psychology has an undeniable impact on the selection of cultural information. If some cultural variants are cognitive attractors (in other words, if they are easier to acquire, store and retrieve than others), then they will be more reliably copied and more likely to be selected than other variants. Psychological mechanisms, in other words, are significant factors in cultural transmission. Intermediate representations might be transformed into cognitive attractors, but cognitive attractors are cultural representations whose content is passed intact between cultural generations.

In summary, even when it is possible for people to select any variation at all from a continuous spectrum of variation, in practice they have to discretize the information in order to discriminate amongst its possible values. In addition, human cognitive processes will ensure that the actual variation in a population will be concentrated around discrete cognitive attractors that are reliably transmitted across cultural generations. Henrich and colleagues (2008) have not shown that discrete representation is not necessary for cultural evolution. Rather, when we look at their work in conjunction with Dietrich and Markman's, we find a convincing argument that even when a cultural population forms a continuum, human cognition will rapidly discretize it, producing chunks of information with persistent cultural heredity. In these ways, we can see how human cognition supports the differential transmission of discretely represented cultural information, thus creating the mechanisms of cultural evolution.

PART II

THE INHERITANCE OF
CULTURAL INFORMATION

Natural Language

4

Natural Language and Culture

The Biological Building Blocks

Information relies for its transmission on receivers that understand the way in which it is represented. If cultural information is represented in the languages of human culture, then its transmission depends on humans who can understand those languages. This chapter explores the innate characteristics that needed to be in place before our species was ready for the most complex, yet also most widely shared, representational system that we know: natural language. It shows how the preadaptations that prepared the way for language evolution were also preadaptations for culture.

I divide human languages into two types. Spoken and sign languages, which have evolved naturally as a means of communication among people, are collectively known as natural language. For languages such as the written word, musical notation or the conventions of architectural drawings, which are realized in objects made or fashioned by humans, I coin the collective phrase "artefactual language." My thesis is that cultural information is represented in a variety of media, in both natural and artefactual languages, and that this information encodes for the phenotypic effects that we call culture.

Humans have a unique capacity for the acquisition and use of language. There are many different natural languages, both ancient (like Latin) and modern (like English), but underlying them all is the same human language faculty. What are its evolutionary origins? Whatever the controversies about the extent to which language is part of our biological endowment, there is a general consensus (Christiansen and Kirby 2003b) that it was preceded by a variety of preadaptations in our hominid past. This means that in order to map the foundations of language, we need first to explore the cognitive, physiological and social capacities of our ancestors (a useful three-pronged analysis, which is due to Hurford [1999]). This chapter sketches what these capacities might have been; shows how each of them would have been crucial not only for the evolution of language but also for the evolution of the culture that language supports; and outlines the biological features that enabled us to acquire a representational system that is both complex enough and ubiquitous enough to act as the substrate for cultural evolution.

Representation

Our capacity for language and culture is founded on cognitive abilities that we share, to a certain extent, with other modern primates. One of the most crucial is the capacity for representation. The evolutionary linguist James Hurford (2003a) points out that some nonhuman primates appear to be able to form concrete and well-defined representations of the world, and to have a degree of voluntary control over these representations: they can attend to an object and form a judgement about it, or evoke the representation of a place and head off for it; and to a limited extent they can problem solve, carrying out mental calculations over their representations.

What do we mean when we say that a creature has formed a discrete representation of the world? There is more than one way to represent a cat, as it were (see Deacon 1997: 69–101). We can use an icon – a type of representation that resembles what it represents – in which case we might draw a picture of the cat, or make a cat-like sound. Another type of representation, an index, will be in some way correlated with its subject matter, as the position of a fuel gauge is with the level of fuel in a vehicle's tank. The most sophisticated type of representation is a symbol, which represents via social convention or established code rather than by resembling or being straightforwardly correlated with that which it represents. For example, the red lamp in a traffic light, when lit, is a symbolic representation of the instruction that the traffic must stop.

Hurford (2003a) has argued that one of the ways in which some animals communicate is by symbolic representation of this kind: via signals that are connected only arbitrarily with their semantic content. He shows how ritualization in animal behaviour, in which a symbolic action is accepted as a surrogate for real action, involves the emancipation of a signal from its original context. When monkeys make their limited use of a variety of alarm calls, for example, we can interpret this as the basis of an ability to make arbitrary associations between signals and meanings, such that perception of the action activates the concept, and attention to the concept may initiate the sound or gesture.

There is no evidence that even very varied animal calls in the wild are learned from conspecifics, and in this sense they are not truly linguistic. Even the Diana monkeys' ability to refine their use of innate alarm calls (see Chapter 2) appears to be the result of a genetically specified learning bias rather than of a capacity that in any way approaches humans' natural ability for the conversational exchange of symbols. Even when captive apes learn a limited set of arbitrary associations between concepts and signals, they use them only to satisfy immediate desires. Nonetheless, although they have nothing like our vast vocabulary, it seems that some animals do have the capacity to learn small sets of arbitrary relationships between signals and meanings, and it is on this "inventory of elementary symbols" that the unique features of human languages are built (Hurford 2003a: 2). The complexities of human language have their evolutionary roots in the much simpler capacity of some other species for discrete symbolic representations.

Imitation

As with representation, so elements of the human capacity for imitation can be seen in other species as well as in humans. Imitation is crucial to language learning and is also, of course, one mechanism of cultural transmission. The capacity for imitation is based on cognitive skills that include an awareness of the creature's own and its companions' activities; the ability to link the two; a degree of means-ends reasoning to tell it why it should want to link them; and a long-term memory sufficient to keep track of the sequence of events that results from all this mental activity (Distin 2005 124–8).

Here again we can see a balance between characteristics that we share with other primates and capacities that are uniquely human. Our instinctive urge to imitate can be observed, to a certain extent, in some other primates. In one experiment (Whiten, Horner and deWaal 2005: 737), when a high-ranking chimp from each of two separate groups was taught one of two different methods of extracting food from the same apparatus, not only did the other chimps in each group go on to imitate the technique of that group's "local expert" but even a "subset of chimpanzees that discovered the alternative method nevertheless went on to match the predominant approach of their companions, showing a conformity bias that is regarded as a hallmark of human culture."

Imitation, though, can take place on more than one level. Although some nonhuman creatures are able to organize sequences of sounds or gestures, they are less good at learning hierarchically structured behaviours (Byrne and Russon 1998). It is one thing to copy the details of a particular behavioural pattern; it is quite another to copy the structural organization of complex processes. To copy a behaviour pattern at this level, a creature needs to be able to pick out which of the pattern's elements are fixed rules and which are variable strategies – in other words, to impose a hierarchical structure on the complexity of sounds or behaviours being copied. The fact that humans are uniquely competent at this level of imitation indicates that we have the cognitive capacity to support the hierarchical transmission of complex information, which is so crucial for both language learning and cultural evolution.

As well as this more complex, structural type of imitation, human children are capable of *selective* imitation: they are able to determine the goal-directedness of observed behaviour, in order to work out why those particular means were used to reach an end. In a seminal experiment by the psychologist Andrew Meltzoff (1988), fourteen-month-old infants observed an adult performing six separate actions, four of which involved objects that were specially constructed for the experiment so that they were entirely novel to all of the infants. One of these novel objects was a wooden box whose top panel was made of translucent orange plastic: the infants observed the adult leaning forward to use her forehead to touch the panel, which was automatically illuminated by a light bulb inside the box. After a one-week delay, two-thirds of these infants imitated her action, which was not produced spontaneously by any of the children in control groups, who had either not seen the light box at all or had observed an adult playing with it and secretly turning the light on, without touching the top panel. Meltzoff (1988: 475) concluded that his results demonstrated the importance of imitation in infant learning, by showing that infants can "internally represent the acts that they see adults perform and are motivated to use these representations to guide their own subsequent behavior, even after the intervention of lengthy delays."

Primates, in contrast, "do not imitate new strategies to achieve goals, relying instead on motor actions already in their repertoire (emulation)" (Gergely, Bekkering and Király 2002: 755) – which in the case of the light box would lead them to use their hands rather than their foreheads to switch on the bulb. Why did most of the human infants not simply reach for the light switch with their hands instead of using the more cumbersome, observed method of leaning forward to use their fore-heads? György Gergely, Harold Bekkering and Ildikó Király (2002) carried out a modified version of Meltzoff's study, in which the adult demonstrated the same forehead action each time, but in one condition the children could clearly see that her hands were free and in another condition they could see that her hands were busy (holding a blanket around her shoulders as if she were cold) while performing the action. What they found was that, although two-thirds of infants in their study, as in Meltzoff's, imitated the forehead action of the demonstrator whose hands had been free, only one-fifth of the infants who had observed the blanket-wrapped demonstrator did the same. Their conclusion was that infants will imitate an adult's means of achieving a goal only if it seems to them to be the most rational option. The children who had observed an adult using her forehead when her hands were busy seemed able to work out that her method would not have been the most rational way of achieving her goal, had her hands been free.

What is interesting about this research is that all of the children who had observed the forehead action did sometimes use their hands to switch on the light: they were all able to emulate the "rational, means-ends aspects" of her behaviour; but what Gergely and colleagues have shown is that the infants were also able to imitate the "nonrational, culturally constrained aspects" of her behaviour (Enfield and Levinson 2006: 17). Even at such a young age, humans perceive that the end result of some actions is not their only goal, which also incorporates the way that those actions are performed. These infants' imitation "is a selective, interpretative process, rather than a simple re-enactment of the means used by a demonstrator" (Gergely et al. 2002).

Research by Josep Call, Malinda Carpenter and Michael Tomasello (2005: 161) supports "the idea that chimpanzees in social learning situations focus primarily on reproducing results (which is consistent with emulation) whereas children focus primarily on reproducing actions (which is consistent with imitation)." Children and chimpanzees were presented with a PVC tube that could be opened in two different ways: a conspecific either demonstrated opening the tube in one of these ways, or demonstrated one of the actions needed to open the tube (without actually opening the tube), or presented a tube that had already been opened in one of the two ways (without demonstrating how it had been opened). The results showed that chimpanzees' actions neither mimicked those of the demonstrators nor differed across information conditions. The children's actions, in contrast, depended on what they had observed, and they even copied actions that had not successfully opened the tube. The chimpanzees, in other words, copied results, whereas the children copied actions. As Andrew Whiten (2005: 54) puts it, "The observational learning of chimpanzees is highly pragmatic, subjugated to individual efforts wherever this gets results. In contrast, children are more prone to copy the actions of others just because others are doing them, betraying an extreme form of reliance on cultural convention."

These experiments confirm, as previous chapters have emphasised, that information depends on a receiver that can detect and react to variations in the source. This is why the same experimental demonstration will bring about a different response in different receivers – and especially in different species of receivers. A fascinating footnote to this research has recently been provided by Friederike Range, Zsófia Viranyi and Ludwig Huber (2007), who have shown that domestic dogs can at the very least give the appearance of engaging in selective imitation. In their experiment, dogs were required to pull down a wooden rod, which opened a food container and hence brought them a food reward. Two groups of observer dogs watched a demonstrator dog pulling the rod with her paw, rather than with her mouth, which a dog would usually prefer to use. For one group of observers, the demonstrator's use of a paw was justified by her carrying a ball in her mouth, but when the other group observed the paw use, there were no constraints that might have explained why the demonstrator did not use her mouth. Afterwards, only the second group of observers used their paws to pull the rod. Does this experiment provide evidence of selective imitation in dogs?

In every case in which a creature observes a demonstrator's behaviour, then infers the demonstrator's goal and works out the means to that goal for itself, there are interesting questions to be asked about the assumptions that the creature brings to its inference of the demonstrator's goal. A human infant, for example, will copy adult actions, almost regardless of the results. In other words, when an infant observes an adult's actions, he assumes that the adult's intended goal is not only the end result but also the action performance: he is motivated to infer action-intentions as well as result-intentions. The evidence regarding chimpanzees, in contrast, appears to be that, although they use their observations of others' actions as clues to how to achieve a result, they do not actually copy them. In other words, when one chimpanzee observes another's actions, it assumes that the other's intended goal is simply the end result and uses the observed behaviour as a clue to how to achieve that result for itself: it seems to be motivated only to infer result-intentions.

What about the dog? Does it simply use the observed behaviour as a clue to how to achieve the result, or does it infer that the other dog's intended goal is the action performance as well as the end result? Does the dog, in other words, assume that the observed behaviour is the product of action-intention as well as of result-intention? We don't know the answer to these questions because, unlike the chimpanzee experiments with the PVC tube, the domestic dog experiments do not include a situation in which the dogs observe behaviour that is unsuccessful in achieving a result. The answer, presumably, will lie in ethological details about whether there is any adaptive value to dogs in acquiring the local "culture," so that they benefit more from doing things in the way that the others in their pack do them than they do from achieving a particular result (a domestic dog's pack includes humans, of course). Clearly, more research is needed.

What is revealed, nonetheless, by the inclusion of nonhuman species in the research to date, is that, on the one hand, the roots of cultural imitation go back a long way, and, on the other hand, humans bring to these situations certain assumptions that are crucial to the transmission of cultural traditions. The natural tendency of young humans to copy cultural conventions, rather than focus on results, is another significant component of the cognitive mechanisms that support both cultural evolution and the transmission of languages – and a "language, of course, is the paramount instance of a set of cultural conventions" (Hurford 2007: 203).

The chimpanzees' actions, in contrast, are unaffected by the actions of the demonstrator. Primates seem incapable of discerning the intentions of those whose behaviour they copy, especially when those intentions are communicative (Tomasello 2006: 516). For humans, the development of a child's interpersonal relationships and place in society seem to hold the key to imitation's importance. Meltzoff and Moore (1994: 96) suggest that a key motivator for early infant imitation in humans is "to probe whether this is a reencounter with a familiar person or an encounter with a stranger." Just as they use physical manipulations like shaking or sucking to probe the function of inanimate objects, so "motor imitation, the behavioral re-enactment of things people do, is a primitive means of understanding and communicating with people" (1994: 83).

Physiology

Physiologically, several characteristics most noticeably distinguish us from other primates. The speed and dexterity of human hand and mouth movements are, of course, the physiological key to our linguistic and tool-using abilities. Humans are unique in the combination of big brains and flexible vocalisation that allowed us to develop a subtler and more complex range of sounds than other creatures. Together with our clever hands, these physiological specialisms enable us to produce the languages – both natural and artefactual – that support cultural evolution.

The size of the modern human brain is not the only way in which it differs from other creatures' brains. Although it does not have any remarkably divergent structures, it is notable for the observable warping of the proportions of its parts, and in particular for the prominence of those parts that handle "verbal short-term memory, combinatorial analysis, and sequential behavioural ability" (Deacon 1992: 64). The human motor cortex is also proportioned differently. Well-known experiments such as those of Wilder Penfield in the mid-twentieth century (Jasper and Penfield 1954) have shown how the areas of the motor cortex assigned to various body parts are proportional not to the size of those parts but to the complexity of the movements that they can perform. What is notable for humans is that the areas that control hands and face are especially large compared with those for the rest of the body. Apes, too, are good at voluntary hand control, but they do not share the specialization in our vocal tracts and our fine control over tongues, cheeks and larynges.

One explanation of the clever hand-vocalisation combination is the aquatic ape hypothesis, according to which we get our clever hands from our primate ancestry, and our clever vocalisation as the by-product of a respiratory system adapted to paddling and diving. The theory is that our ancestral split from other primates, around 6 million years ago, was precipitated by a period in our evolutionary lives during which we spent much of the time swimming, diving and eating aquatic food; and our vocalising abilities were founded on the airway control that evolved initially for swimming and diving (Verhaegen 1988). The aquatic ape hypothesis remains controversial, but some explanation is needed for humans' unique respiratory and vocalisation control.

The fossil record of our species is, of course, complex and incomplete. There is evidence of flaked stone tools from around 2.5 million years ago, the product of our ancestors' clever primate hands, and hand axes first appear around 1.4 million years ago. Hominid specimens from that era have been found with brains of 900–1000 cc, well on the way from the 400 cc of early australopithecines to the 1200–1500 cc of modern humans – a brain size that emerged between six hundred thousand and two hundred thousand years ago in the absence of any changes in body size – and

Steven Mithen (2000: 207–8) argues that this expansion is probably related to the evolution of a language capacity. Robin Dunbar's research suggests that the timing of language evolution does fit this hypothesis. Dunbar (2003: 175–6) draws together various strands of evidence in support of the claim that some form of language must have evolved with the appearance of *Homo sapiens*: this includes evidence from his own studies of grooming-time requirements and from anthropological and paleontological research into the sizes of the hypoglossal and thoracic vertebral canals.

It is worth bearing in mind that there is as yet no real consensus about the evolutionary relationship – cause, concomitant or effect – between each of our specifically human adaptations: our advanced manual, vocal, social and cognitive abilities. There is also some disagreement about the reason primates, and especially humans, should have larger brains than other species. The traditional view has been that primate ecological strategies required more complex problem solving than those of other species. This view is largely based on observations about primates' frugivorous diet, which requires them to range more widely than many other species, meaning that they need a better memory in order to hold information about the large-scale, uneven distribution of their food. Moreover, much of this food cannot be accessed without extractive foraging techniques (termite dipping, removing fruit cases, hunting and so on), which are dependent on relatively advanced mental capacities. More recently, however, it has been suggested that primates' complex social systems make computational demands that select for increases in brain size, and specifically for increases in the neocortex (Dunbar 1998: 178–9; 2003).

Sociality and Cooperation

Just as the anatomy of our brains shows both similarities to and differences from the anatomy of other primates' brains, so our lives as social creatures are founded on characteristics that to a certain extent we share with other modern primates. Many apes also live socially, and they have some – though to what extent is still disputed – limited mind reading (i.e., awareness of others' mental states) and social manipulation skills (Byrne and Whiten 1988), as well as the apparatus for establishing joint attention with conspecifics (Emery et al. 1997). But although many animals manipulate social situations to a certain extent, humans are uniquely good at it. Other creatures don't seem interested in communicating about anything other than their own immediate needs. Even the somewhat contentious chimpanzee theory of mind has nowhere near the complexity of ours.

What has given rise to this difference in the levels of sociality of humans and other apes, which according to the social brain hypothesis has been crucial for the emergence of our big brains? Hurford (2007: 187–8) points out that social group sizes are strongly affected by ecological conditions like food availability, diet and the frequency of predation; by the amounts of time spent on grooming and vocal communication; and by both intragroup and intergroup social structures. He draws together evidence (2007: 194–7) that the human evolutionary trajectory has probably been away from the gorilla-like, male-dominated harem arrangements of australopithecines, towards social arrangements that incorporate both long-term monogamy (which is in some respects gibbon-like) and a more fluid, chimpanzee-like fission-fusion structure, with intermittent movement of individuals between groups. He notes that although bonobos are the primates that have been most disposed to developing human-like communication in captivity, and gibbons have been observed to make the most complex vocal displays, both fall far short of the communicative abilities of humans.

"It is important to appreciate in this context," cautions Dunbar (2003: 164), "that the contrast between the social and more traditional ecological/technological hypotheses is not a question of whether or not ecology influences behavior, but rather is one of whether ecological/survival problems are solved explicitly by individuals acting on their own or by individuals effecting social (e.g., cooperative) solutions to these problems. In both cases, the driving force of selection derives from ecology, but the solution (the animals' response to the problem) arises from contrasting sources with very different cognitive demands (individual skills in one case, social-cognitive skills in the other)." The fact that humans tend to effect cooperative solutions to ecological problems provides a fertile arena for the transmission and evolution of cultural information.

Indeed, cooperation is a key aspect of human sociality. The very use of a shared language is a form of cooperation, in which participants assent to the use of a conventional communicative code (Hurford 2007: 270). In the absence of a cooperative disposition, language use is neither possible nor, indeed, advantageous: as Hurford points out, enhanced communication is biologically advantageous only to species in which cooperation is a low-risk activity. Several strands of psychology indicate that humans are instinctively motivated to cooperate with one another (Enfield and Levinson 2006: 10–11) and are consequently disposed to share information. It is obvious that such a disposition would provide an engine for cultural transmission.

Species whose social lives are essentially competitive, rather than cooperative, do not engage in even the most basic forms of communication with one another. Humans depart from the rest of the apes, in communicative terms, before natural language has even appeared over the horizon. Michael Tomasello and colleagues (Tomasello, Carpenter, & Liszkowski 2007: 720) suggest that pointing behaviour in preverbal infants is "an existence proof that human-style co-operative communication does not depend on language, and suggests rather that language depends on it." Human communication is a collaborative, public process of meaning construction, characterized by systems of rapid turn taking and immediate repair of any misunderstandings (Goodwin 2006), which make it possible for humans to interact even without conventional language, with the help of resources like "eye gaze, facial expression, body movement, phrasal breaks, repair, and the environmental surround" (Schumann et al. 2006: 16).

It is for this reason that Tomasello (2006: 520) urges researchers to postpone asking why apes do not have language and focus instead on asking why apes do not even point. He cites research which demonstrates that the answer is certainly not that apes are unable either to use gestures or to follow conspecifics' gaze direction with sufficient flexibility, and he asks why, then, apes do not sometimes use anything like a pointing gesture to direct one another's attention, for example to something that another ape cannot see (2006: 507). Why do they not even comprehend what is meant when a human uses such a gesture for their benefit? His answer is that, in contrast to human infants as young as fourteen months, apes are generally unable to use communicative cues: they fail to understand that the human's pointing or looking behaviour is informative and done for their benefit (2006: 508–9). What apes do not seem to establish with one another (or even with human trainers, when in captivity), and humans do, is a "joint attentional frame, or common communicative ground" (2006: 509).

Human infants display several different motivations for their use of pointing gestures. They point imperatively (when they want something), declaratively (when they want an adult to look at something) and sometimes even informatively (when they want to give an adult information about something). Even in captivity, however, when apes' communicative potential is elicited by a more cooperative environment than they experience in the wild (Hurford 2007: 239–40), apes can sometimes

learn to point imperatively, but they never point declaratively or informatively. This, according to Tomasello (2006: 513), is because they are
neither motivated to help others by behaving informatively or sharing
experiences, nor indeed aware of what information others do not yet
have. Children as young as twelve months, in contrast, show by pointing
informatively that they have some awareness of another person's lack of
knowledge, as well as an early instinct for cooperation (Liszkowski 2006).

Tomasello's conclusion (2006: 515–6) is that chimpanzees use and
understand gestures as functional, one-way mechanisms for achieving
an aim, whereas humans understand communicative conventions like
pointing and language as essentially collaborative acts, in which communicators adjust their actions in accordance with recipients' signals of
(in)comprehension. In evolutionary terms, Tomasello's proposal (2006:
517) is that individual humans who excelled in collaborating with others
were at an adaptive advantage and that, as a result of their emergent
sociocognitive skills, humans became uniquely able to collaborate with
others, with whom they could form and jointly commit themselves to
shared goals.

Theory of Mind

According to Meltzoff, it is this immersion in social interaction with
other humans that is responsible for children's acquisition of a theory of
mind. He presents evidence that human babies are born with an innate
tendency to imitate (Meltzoff 2005: 77), the ability to represent their
own acts and other people's supramodally (2007: 126) and an innate
tendency to believe that other people are "like me" (2005: 77). There
is a basic equivalence, in young infants' minds, between themselves and
other people, and this gives them a framework for interpreting and
understanding observed behaviours (2007: 126). The result of their
immersion in social interactions with other psychological agents whom
they perceive in this way is an adult theory of mind.

Without a theory of mind, we can neither model others' inner states
nor craft our communicative actions in just the right way to ensure that
others will understand our intentions (Enfield and Levinson 2006: 11).
A theory of mind, in other words, is one way of honing the accuracy of
cultural transmission. But Jennie Pyers (2006) has shown how language,
in turn, is crucial for the acquisition and application of a theory of
mind. Pyers reports on the emergence of Nicaraguan Sign Language
and the difference in theory of mind capacities between younger and

older signers. People who came late in life to what was, in any case, initially a pidgin, were unable to deal with standard false belief tasks: there seemed to be a causal connection between their limited language skills and an impaired theory of mind. People who had been exposed to a more developed sign language from a younger age, on the other hand, had a greater capacity for theory of mind. There would appear to be a positive feedback loop between the emergence of better communication systems and a theory of mind. The correlation between language and shared intentionality, or theory of mind, is supported by evidence from autism: high-functioning autists and people with Asperger syndrome can use language in a very literal, nonmetaphorical way even though their theory of mind is impaired, whereas more severely autistic individuals have both a severely impaired theory of mind and little or no language (Hurford 2007: 319).

Cultural Preadaptations

Some two hundred thousand years ago there emerged a species that combined finely controlled breathing and hand and mouth movements with a large and uniquely proportioned brain. The physiological tools of language and culture were in place, and the same species was capable of imitation and of forming discrete representations of information. Crucially, members of this species were also innately disposed to share information with one another. They favoured cooperative solutions to ecological problems, and their communicative skills were honed by an innate tendency to discern and reveal intentions as well as actions. Members of this species lived in unprecedentedly large social groups, and they had an innate tendency to imitate the social conventions of the group rather than merely to use conspecifics' behaviour as a clue to how they could survive as individuals.

In short, the evolutionary foundations of natural language were also the foundations of evolution in cultural information, for here we had a species that could represent information, which members tended to share with one another, in increasing amounts, with increasing accuracy, selectively favouring existing cultural conventions over their variants.

5

How Did Natural Language Evolve?

If cultural evolution is a separate process from biological evolution then it must be the product of a separate inheritance mechanism. I have shown that inheritance in any realm is dependent on generations of receivers that can detect and react to variations in that which is inherited. This means that cultural evolution can happen only in species whose members can detect and react to cultural variations. The previous chapter outlined how the human species was preadapted, or in other words biologically prepared, for culture. In particular, our tendencies to live in large social groups, to share information with one another, and to notice and imitate social conventions are all key factors in cultural evolution. But it was not until natural language became widespread among our ancestors that cultural information had a mechanism for inheritance. The emergence of a shared natural language created generation after generation of receivers for the cultural information that the language carried. What cultural evolutionary theory needs, then, is a Darwinian account of the evolution of natural language in our preadapted species.

Against the background of a range of cognitive, social and physiological preadaptations such as those addressed in the previous chapter, it is not hard to understand why an emerging capacity for precise and complex communication would have enabled us not only to survive in a broader range of environments but also to live in larger and more complex social groups. As a mechanism to bind social groups, language makes much more efficient use of bonding time than physical grooming does: it means that we can "groom" more than one person at once, that we can learn and inform others about what has happened in our shared social network when one of us was absent and that we can continue to "groom" while walking or feeding (Dunbar 2003: 174).

But the intuitive advantages of skilled communication do not answer all of the evolutionary questions about the emergence of natural language: "the central puzzle is the relation between intricate universal principles of grammatical structure and fitness" (Hurford 1999: 6). Humans have a vast learned vocabulary, the ability to use learned rules for the complex composition of long sentences, and a rich conceptual system. The modern level of transmission of cultural information is possible only because all of these components are widespread among humans. What we need, therefore, is a Darwinian account of how such apparently gratuitous complexity could have increased our fitness.

James Hurford's immensely persuasive thesis is that the structural complexity of natural language increased our ancestors' communicative abilities, and it was this that increased their fitness. The adaptive advantage of communication was so great for this uniquely intelligent, cooperative[1] and representational species, that it provided the evolutionary impetus for the intricate grammatical structure of modern natural language. "The great gift of language is not only that it enables us to think, for there are ways of thinking that do not require language, but that it enables us to communicate to one another and across generations" (Price and Shaw 1998: 189).

Instinctive Communication

Research by György Gergely and Gergely Csibra (2005) confirms that human adults have an instinct to communicate with children and that human infants have an instinct to act on adults' ostensive communicative cues. In a further modification of the light box experiments described in the previous chapter, the infants were divided into two groups. One group was further subdivided and presented with the forehead action, hands free or hands busy as before. The other group was subdivided in the same way "but without being exposed to any ostensive-communicative cues by the model" (2005: 477). The usual methods for studies of infant imitative learning include attracting the infants' attention by means of speech, eye contact, name use and pointing or looking at the object

---

[1] Note that terms like "intelligent" and "cooperative" are shorn of any approbation in this context. They accurately describe human behaviour relative to the behaviour of other species, even though they do not always accurately describe human behaviour relative to the normative standards of rationality and cooperation that we often wish to apply in a social context.

(2005: 473). The results from the group with whom these cues were used in the normal way replicated the previous results of the light box studies. But when the demonstrator's actions were "incidentally observed" by the infants, rather than cued in the normal way, although some of the infants in both subdivisions of this group did imitate the forehead action, there was no difference between the sorts of imitation evoked in the hands-free and hands-occupied conditions (2005: 477). It seems that the infants' capacity for selective imitation was dependent on ostensive-communicative cues from the adult.

Gergely and Csibra (2005: 472) conclude that "humans possessing cultural knowledge are naturally inclined not only to *use*, but also to *ostensively manifest* their knowledge to (and for the benefit of) naïve conspecifics," and such learners are "predisposed to interpret the teacher's ostensive communicative cues . . . as evidence that the manifestation will contain *new* and *relevant* cultural information."

Critical Period

Given that our species has such a strong instinct for communication, what adaptive benefit could our ancestors have derived from the emergence of a communication method like natural language? One answer is that the local language helped to identify the conspecifics with whom it was safe to cooperate. Just as communication is advantageous only within societies of cooperators, so "cooperation is likely to be viable only where there is relatedness or guaranteed reciprocity between individuals" (Nettle and Dunbar 1997: 93). Language, in this context, appears to be functioning as a badge of group identity: a way of extending the size of cooperating groups beyond the ties of known kinship. An important aspect of children's language learning is the dialect and accent of local native speakers (Nettle and Dunbar 1997): experiments with infants and young children suggest that even prelinguistic infants prefer people who speak with the local accent, and that young children favour otherwise-unfamiliar people simply because they use the local language (Kinzler, Dupoux and Spelke 2007).

This is significant because a multitude of sociolinguistic research has demonstrated the extent to which the right linguistic markers can determine access to cooperation. For example, when people are faced with aggression or hostility they tend to broaden their accents and emphasise their use of local dialect or language, whereas in positive interactions speakers will automatically accommodate their ways of speaking to one another (Bourhis and Giles 1977; Giles and Smith 1979; Labov 1963; LePage 1968). Experiments have shown that, when asked to assess speakers' friendliness and helpfulness, we tend to rate our own local speech varieties higher than those of others (Giles and Powesland 1975). Research has also verified both the common observation that people often adapt their speech patterns to conform more closely to prevailing expectations and the fact that such adaptation does indeed bring social and economic benefits (Nettle and Dunbar 1997: 94–5).

Studies on the impact of learning more than two languages confirm the ways in which a native language is persistently perceived as "ours" and all subsequent languages as "foreign." Research indicates, for example, the perhaps surprising result that a second language will have more influence than a native language on the ways in which speakers compensate for deficiencies in their third language. The reason appears to be that speakers with deficits in their knowledge of a third language prefer to borrow from another language that they perceive as "foreign" rather than from their "own" native language (De Angelis and Selinker 2001). If the native language is so deeply felt as a badge of group identity, then Hurford (2007) concludes that there must be an adaptive advantage in learning society's collective communication system before mating age.

Human children have a much longer period of dependency on their parents than other primates, and this may well be "related to bipedalism, with the narrowing of the pelvis, and the consequent adaptation to be born with a smaller head. Brain development was postponed in human evolution to a postnatal stage, opening the door to a far greater influence from the environment, i.e. to learning" (Hurford 2007: 196). The early, extended opportunity to learn communication skills, which is provided by such a long period of dependency on adults, appears to be crucial to language learning. There is a "critical period" for language acquisition, whereby although the ability to learn new languages persists into adulthood, the "ability to speak a language as a 'native' seems to be restricted to a very brief period in early childhood" (Dunbar 1999: 203). Human children are innately disposed to take adult noises seriously – they take on trust the fact that the adult noisemakers around them are providing information that is genuinely useful and truthful (Hurford 2004: 3) – and they spend years doing little other than learning how to make the same sorts of noises. Clearly, this is also crucial for cultural evolution, for it reveals that we are genetically disposed to learn the representational system that will enable us to receive the information that other individuals are trying to communicate, and that we have an extended opportunity, from our earliest days, for its acquisition.

Communication versus Interpretation

Natural language has some interesting structural features, which support the theory that it evolved for *communicative* rather than for *interpretative* success. In other words, natural languages are structurally better suited to expressing a speaker's thoughts than they are to ensuring that the listener can unambiguously understand what is said. The distinction between communication and interpretation becomes apparent when we think about how quick we are, in ordinary conversation, to blame the other person for misinterpreting our meaning, before it occurs to us that perhaps we did not make ourselves clear.

In structural terms, Hurford (2003b) draws our attention to the widely accepted fact that natural languages seem innately disposed to avoid synonyms (different words that carry the same meaning), but abound in homonyms (words that look or sound the same but carry different meanings[2]), despite the confusion that homonyms can cause. Hurford has used computer simulations of innate dispositions to synonymy and homonymy to explore why this should be the case. What his models show is that when languages evolve primarily for the purpose of speakers being able to communicate with one another, those languages tend to avoid synonyms: they evolve to produce only one signal from each given meaning. In contrast, when languages evolve primarily for the purpose of hearers being able to interpret the incoming message correctly, those languages tend to avoid homonyms: they evolve to facilitate the inference of just one meaning from each given signal. An innate human preference for homonymy over synonymy seems, therefore, to imply that human languages evolved primarily for the production of signals rather than for the reception of precise meaning. "We may be primarily speakers, and secondarily listeners" (Hurford 2003b: 450).

It might seem that there is something counterintuitive about the claim that natural language evolved for communication *rather than* for interpretation: surely couldn't we have one without the other? To a certain

---

[2] More precisely, David Rothwell's entertaining *Dictionary of Homonyms* defines a homonym as "a conceptual word that embraces both homographs and homophones. Hence 'homonym' can refer to a word that is spelt the same way as another word but sounds differently, to a word that is spelt the same way and sounds identical too, and to a word that sounds the same as another word but is spelt differently" (Rothwell 2007:8).

extent this is true, of course, but we have seen that what matters, if information is to be disseminated among a group of receivers, is that they should all be proficient in the system in which it is represented. In other words, what matters for communication between these receivers is that they should share a representational system. Ideally, of course, they would share a perfect representational system, but we know that evolution has no foresight: selective pressures produce good-enough results, not perfection. What Hurford's computational analysis has shown is that natural language evolved primarily as a method of sharing information. The result was a system that enabled language users to receive a great deal of information from one another, even if sometimes the system let down the receivers and they needed to rely on extralinguistic techniques for clarification. It might not have been perfect, but it was certainly good enough to facilitate the evolution of cultural information.

It is interesting to note that, physiologically, this view is supported by the evidence from the proportions of our motor cortex, which are biased so much towards speech over hearing. It is also supported by evidence that human conversations are deeply collaborative (Goodwin 2006): they involve the ongoing possibilities of both self-repair (when we immediately correct a word or phrase that we realise might be misinterpreted) and checking the meaning of what the other person has just said (Schumann et al. 2006). This collaborative process of meaning construction relieves spoken language of the need for absolute interpretative accuracy.

Communication versus Representation

Natural language also has structural features which imply that it evolved for communicative rather than for *representational* success, or in other words, that the primary purpose of speech is to express our thought to other people rather than to represent it to ourselves. The spoken language conforms to certain rules about how its units are put together, many of which are irrelevant until we are trying to express our thoughts in sounds. Hurford (2002) surveys a range of morphological and syntactical structures, none of which plays a purely representational role. It is not until you try to communicate your thoughts that you need phonology or morphology, or indeed many of the main devices of syntax: rules about how sounds are put together to form syllables and units of meaning, or rules about how these elements are put together to make sentences. This is not to say that the spoken language does not represent information at all. The claim is, rather, that language evolved primarily for the audible expression of information to other people, and not primarily for the accurate representation of information to oneself.

It is, of course, an inescapable characteristic of representational systems that their structures reflect their media. The structure of any method of representing information, whether in our heads or in our speech, in our gestures and facial expressions or on a page, is inevitably determined by the nature of its medium. The way in which the medium is structured *is* the representational system, and if you want to put the same information into a different medium then it will need to be structured differently. You simply cannot structure spoken sounds in the same way that you can structure thoughts or hand gestures or written symbols. Those media do not work in that way.

This means that the answer to the question, "Why does natural language have a sound-based structure, rather than a structure that reflects the content of our thoughts?" is at one level simply, "Because it is made up of sounds, and it could not convey meaning unless it had a structure that works for sounds." This is an answer that tells us nothing about the reason for which natural language evolved – but it does require us to rephrase the original question. If natural language evolved as a method of representing information in a sound-based medium, then instead of asking why that system has a sound-based structure (answer: because it uses a sound-based medium), we need to ask why there should have been pressure towards using sound as a medium for representation in the first place. Why would we bother with an alternative means of representing the content of our thoughts – especially with one that potentially requires them to be packaged in a different way from how they are structured internally?

The most obvious characteristic of sound is that it can be heard. Of course it can be heard by the speaker as well as by the listener, but it is not the only way in which the speaker can access his own thoughts, whereas expressed language *is* the only way in which somebody else can access many of the speaker's thoughts. Thoughts can be expressed in signs as well as in speech, but sound has the advantages that it is hands free, that it carries through the dark and through thick forest, and that it does not require the hearer to be looking at the speaker for the signal to be understood (Hurford 2007: 185). This is not conclusive evidence that speech evolved for communication rather than as support for thought. It remains theoretically possible that the evolution of a shared medium of communication was adaptively advantageous because it enhanced our individual cognitive abilities rather than because it enhanced our ability to communicate with one another. But it is at this point that Hurford's other arguments kick in.

The evidence that speech evolved because of the adaptive advantages of being able to communicate with other members of your community, rather than as a means of enhancing your own representational skills, comes from a variety of sources to which he and others have drawn our attention. The sounds that we produce are not intrinsically meaningful: they must be marshalled by the rules of natural language if they are to carry any information, and there is a critical, prepubescent period during which we are innately disposed to acquire these rules. If their function were primarily to support internal cognition, then it is less obvious that there should have been an evolutionary incentive for us to ensure that we acquire them before we try to acquire a mate, or indeed for us to use the same words as other people use in our society, even when talking to ourselves (Hurford 2007: 174). Nor are these rules ideally suited to the precise representation of our thoughts. Natural language has a structural bias towards context-dependent disambiguation rather than strict unambiguity (it prefers homonyms to synonyms), which implies that it is more concerned with the immediate production of a signal in a system that other receivers share, than it is with the accurate representation of content.

What matters for cultural evolution is that the evolution of natural language provided a shared representational system, which enabled users to acquire information from one another. It was not a representationally perfect system, and in time this would prove problematic. In the meantime, a cooperative species, with a biological incentive to acquire information from one another, and an emerging system for its transmission, provided the substrate for the beginnings of cultural evolution. The selective advantage that natural language gave humans, by increasing their ability to communicate with one another, was paralleled by the selective advantage that it gave cultural information, by increasing its ability to be transmitted between humans.

Language Evolution: Biological or Cultural?

There is good evidence, then, that language evolution was propelled by the biological advantages of communication for a naturally cooperative species. This leaves open the questions to what extent language is an evolved biological adaptation, and to what extent it is culturally transmitted.

Clearly, it shares elements of both. There is a broad consensus that language acquisition is innate, but it is equally obvious that the differences between languages do not correspond with genetic differences between their speakers: there are learned variations in language lexicons, syntax and phonology. Most would also agree that natural language use is governed by universal structural principles, but the fact that certain language universals emerge from the analysis of many languages, or from the comparison of historically distant languages (Kirby 2007: 3), need not imply that these universals are innate. Although nobody would deny that the human language faculty is both enabled and constrained by our biological endowment, its universal features could be the result of universal properties of language use (Kirby 2007: 3). So, whereas for evolutionary psychologists like Steven Pinker these universal principles are dictated by innate, biologically evolved mental structures, for evolutionary linguists like James Hurford or Simon Kirby they are largely the result of evolution in the languages themselves.

The linguist Noam Chomsky is the best-known and most influential proponent of the view that language universals are innate. He has concluded that human infants have an innate knowledge of the universal grammar that will govern whichever language they experience: a biological endowment ensures that they can acquire language easily and quickly, and that human languages will all develop along mutually comprehensible lines. Perhaps counterintuitively, however, he has denied that our innate language acquisition device can be explained by Darwinian natural selection, and he was supported in this view by the evolutionary biologist Stephen Jay Gould, among others (e.g., Piattelli-Palmarini 1989). "Chomsky and Gould have suggested that language may have evolved as the by-product of selection for other abilities or as a consequence of as-yet unknown laws of growth and form. Others have argued that a biological specialization for grammar is incompatible with every tenet of Darwinian theory – that it shows no genetic variation, could not exist in any intermediate forms, confers no selective advantage, and would require more evolutionary time and genomic space than is available" (Pinker and Bloom 1990: 707).

In contrast, Steven Pinker and Paul Bloom (1990) have concluded that "there is every reason to believe that a specialization for grammar evolved by a conventional neo-Darwinian process." Just as for other complex systems, like the eye (Pinker 2003), they believe that the best explanation for the appearance of complex design in human language is biological adaptation. The capacity to gather and exchange information brings great adaptive advantages: it enables us to tap into knowledge that other people have accumulated, and to defend ourselves against environmental changes or threats much more speedily than the normal processes of evolution would allow. In the light of these adaptive advantages, Pinker and Bloom (1990) see the human language faculty as an evolved specialization "for the communication of propositional structures over a serial channel." In other words, the human language faculty has evolved so that we are innately able to encode the information in our private thoughts (Pinker and Bloom's "propositional structures") in a format (spoken sentences) that can be gathered and exchanged by other people – and this instinct for language acquisition is a biological adaptation.

There is, though, a Chomskian foundation to Pinker's language instinct: he would agree that we learn to encode information in accordance with the rules of an innate mental programme, the universal grammar, which explains the speed and ease with which infants acquire language. There are many who would now dissent from this hypothesis. In a report on the Language Evolution Roundtable held at the University of California, Los Angeles, in November 2004, for example, the linguist John Schumann draws together a synthesis of the views of the many researchers in this field who took part in the roundtable, and concludes that the "structure of language as manifest in grammar may be a complex adaptive system that emerged as a cultural artifact and that exists between and among brains" (Schumann et al. 2006: 26).

In a similar vein, Stephen Levinson (2003: 27–8) argues that "the biological endowment for language must be . . . a learning mechanism wonderfully adapted to discerning the variability of culturally distinctive systems – a mechanism that simultaneously puts limits on the variation that those systems can throw at it. On this account, the essential properties of language are divided between two inheritance systems, biological and cultural, and the long-term interactions between them."

Patricia Kuhl and Andrew Meltzoff (1997: 8) agree that, although infants have innate abilities that are "highly conducive to the development of language," it is "linguistic experience" that "fully restructures the system." On their view, infants are born with innate "auditory boundaries" (1997: 35), which have strongly influenced the selection of sounds for use in the various natural languages, but are functionally erased by exposure to ambient language. By the age of six months, infants' speech perception has already been altered by language exposure (Kuhl et al. 1992), and by twelve months, infants can no longer discriminate contrasts in foreign languages, which previously they were able to discriminate

(Werker and Lalonde 1988). Unsurprisingly, Kuhl and Meltzoff (1997) conclude that language acquisition is not solely determined by infants' innate abilities.

Hurford is another of those who agrees with Pinker that humans' unique capacity for language acquisition is evolved, but disagrees with Pinker's assessment of the role that cultural transmission has played in language evolution. Hurford's thesis is that the primary impetus for language evolution was communicative cooperation – or, in other words, so that all members of a group could use the same communicative code – and that once humans were language ready, as the result of a collection of biological preadaptations, linguistic complexity would have increased via an intergenerational process of social learning.

In *The Selfish Meme* (Distin 2005: 161–7) I supported the opposing view – that we have a language instinct underpinned by an innate universal grammar – and argued that it was compatible with my account of language as the DNA of culture. I am not persuaded that the two are totally incompatible, but I was not at the time familiar with this more balanced account, which, as will become clear from what follows, provides by far the more satisfying, coherent and convincing explanation of both linguistic and cultural evolution.

Children are not explicitly taught language in the way that they are later taught to read and write. They just pick it up from what is being spoken all around them: an extremely impoverished input whose content will vary enormously between individuals. Despite the apparent inadequacy of this input, children acquire intricate linguistic abilities with remarkable speed. Despite the variation among the input, their resulting language is (in all important respects) uniform. One of the attractions of a Chomskian account is that if our early language acquisition is innate, then this explains not only the speed and ease with which we pick up language, but also the uniformity. Given limited English input, our language instinct will enable us easily and swiftly to produce English output, and only English output. In providing us with a way to systematize the input, the universal grammar also limits us to that way.

Hurford's thesis, however, is that the underlying grammatical uniformity of natural languages is the result of coevolution between these languages and the human brains that are required to learn them. Hurford (2004: 1) points out that there "are two features of human language (including manual sign languages) that are simply absent from the natural communication systems of any other species. One is learned arbitrary symbols, and the other is recursive, semantically compositional, syntax."

The explanatory value of his thesis becomes apparent once we ask how our communication system evolved from basic symbol use to the compositional syntax of modern natural language. He argues that complex compositionality did not emerge *despite* the variability and poverty of the linguistic input to which children are exposed – but because of it. The significance of Hurford's thesis for cultural evolution is that it explains how natural language, the means by which we interpret much cultural information, is itself a cultural acquisition. This provides further support for Chapter 3's conclusion that cultural evolution is not simply an additional dimension of Darwinian evolution, because neither cultural information nor its means of interpretation is part of our genetic endowment.

Compositionality

What do we mean by a "compositional" language, and what would a noncompositional language look like? The meaning of a portion of compositional language is derived from the meanings of its constituent parts and the way in which they are put together. Compositional phrases like "the bees are buzzing" or "the dog is brown," for example, can be understood by anyone who knows what their constituent words mean and how English puts words together. Modern natural languages are on the whole distinguished by their compositionality.

What compositional languages compose are "listemes." A listeme, whether it is the size of a small word or of a longer phrase, is "an element of language that must be memorized because its sound or meaning does not conform to some general rule" (Pinker 1994: 478). For instance, colloquial phrases like "the bee's knees" or "a dog's dinner" make sense only as a whole. Knowing the meaning of *bee* and *knee*, or *dog* and *dinner*, together with the rules of English grammar, would not allow you to work out the meaning of the these phrases if you were not already familiar with them. Noncompositional languages comprise only listemes and no rules. In noncompositional languages, only the total signal carries any meaning: its meaning cannot be derived from the meanings of its individual parts.

What this means is that a language user who is exposed to a noncompositional language can competently express only meanings that he has already observed. Because there are no compositional regularities in the language, the language learner cannot make generalizations from the sample of the language to which he has been exposed. It follows that novel meanings can be expressed only through invention – and that if too many novel meanings are generated, so that the language's vocabulary grows sufficiently large, then in effect the language will not be learnable.

In contrast, for a language user exposed to a compositional language, expressivity is proportional not to the number of utterances that he has observed but to the number of language features that he has observed (the arguments in this paragraph and the next are taken from Brighton and Kirby [2001]). If there is a learning bottleneck, such that learners are exposed to only a small subset of possible meanings during their lifetime, then a language needs to be compositional in order to be learnable. Compositionality is of little use, on the other hand, if there is no such bottleneck, because learners are exposed to a large proportion of the language during their lifetime.

In evolutionary terms, this means that "in a population capable of both rote-learning and acquisition of rules generalizing over recurrent patterns in form-meaning mapping, a pressure exists toward an eventual emergent language that expresses meanings compositionally" (Hurford 2003a: 8). Words gain novel grammatical roles through frequent use in a particular context, and the new rules subsequently govern the language use of future generations. The rules governing the previous generation's language use will be interpreted and internalized by the current generation – and the current generation's language use will, in turn, form the raw data on which the next generation's interpretation and internalization of language rules will be based. As the number of shared symbols increases, so there develops an intergenerational bottleneck in their transmission: of all the possible symbols, only a small subset is observed by new learners. This bottleneck forces the emergence of a compositional language: the poverty of stimulus, an effective bottleneck through which languages are passed from one generation to the next, is actually necessary for the emergence of a regular compositional language.

Hurford and Kirby (1999: 2) have also argued that the critical period for language acquisition has coevolved with the *size* of human languages. "Put simply, the speed at which an individual can learn the language of the community, plus a critical period in which it can be learnt (both biologically given), together determine the maximum size of the language the individual can command as an adult. As this is true for all individuals, a limit on the size of the language as it exists in the community, and the typical age-span in which it can be learnt, are determined by these biological factors." Their model "depicts a self-feeding spiral of language size responding to increases in speed of acquisition, and speed of acquisition in turn responding to increased language size" (1999: 19).

Conclusions

A lot of ground has been covered in this chapter and the last, and it may be helpful at this point to summarise their conclusions. First, there is broad agreement that the human language capacity evolved against a background of preadaptations including representational abilities, physiological changes and sociality. In particular, the previous chapter drew together research from linguistics, psychology and behavioural ecology, which indicates that the roots of natural language lie in a uniquely human instinct for cooperation. Our ancestors solved their ecological problems cooperatively, using increasingly advanced sociocognitive skills. Other primates can learn from one another, but only humans are capable of sophisticated structural and selective imitation, which takes into account the organization and motivation of learned behaviour. Human infants show a conformity bias that far outstrips other primates' desire to be like one another. Children imitate culturally arbitrary elements of behaviour as well as rational, means-ends actions, and they spend the first years of their lives doing little other than absorbing the linguistic and cultural norms of the society into which they are born. Surrounded from birth by social agents who are themselves naturally inclined to share information, the human infant has a supramodal instinct to perceive these adults as being "like me": to imitate them and begin to understand that they have similar mental processes to his own. Language acquisition, which could not occur without this instinct for conformity, subsequently facilitates the development of a more advanced theory of mind, in a positive feedback loop between communication and commonsense psychology.

Without this instinct for cooperation and conformity, children could not acquire language, and without access to the local system of communication they could not find mates when they are older. By enhancing communication, language acts both as a conduit for cultural norms and as the social glue that holds large groups together. One of the ways in which language enhances sociality is by acting as a marker of group identity. Children's brains are permanently altered when their early linguistic experiences tailor their perception of language for the local variant, and this will have an influence not only on their use and understanding of language but also on their inclination to cooperate with others whose speech patterns they perceive as (un)familiar. Our instinct for cooperation, on which the benefits of enhanced communication rely, is given direction by the ways in which we learn to communicate.

In a species that relied on social solutions to ecological problems, the evolution of language was propelled by the selective advantages of enhanced communication among those who are disposed to cooperate with one another. In order to understand what a speaker means, a listener must grasp not only the meaning of his words but also "what is implicated in a given context assuming that the speaker intends to be cooperative" (Stanovich and West 2003: 4). Local language variants are one of the ways in which we identify those who are disposed to cooperate with us. The evidence points to the cultural evolution of language, on the back of the genetic advantages of enhanced communication for members of a species with an instinctive conformity bias. As humans cooperated with one another in sharing information, socially agreed rules emerged for the expression of meaning. The intergenerational bottleneck, through which an increasing number of shared symbols must pass, guaranteed the emergence of compositional rules for their interpretation. There are features of modern language structure which indicate that its primary purpose is communication rather than interpretation, and a communicative imperative also provides the best explanation of its basis in a sound-based medium. The spoken language is a cooperative game played according to socially determined rules, which are learned during a biologically critical period, and nonplayers are denied access to the folk psychology, cultural norms and local identity badge that they need for biological success.

The biological advantages of enhanced communication should not, however, be taken as evidence that natural language is a biological artefact. Rather, natural languages are cultural artefacts that coevolved with the human brain, thriving in an environment in which humans were biologically rewarded for becoming better communicators. Language-using early humans had the novel ability, uniquely among species, to share all of their existing knowledge and to cooperate on shared tasks. Our ancestors were increasingly able to give, receive and manipulate information. Parental and tribal traditions could be passed on explicitly as well as by imitation of adult behaviour, whether they involved beliefs, skills or attitudes. By cooperating with one another, sharing their knowledge and talking about what they were doing, these early humans developed a culture that expanded in a way that was impossible before the dawn of natural language.

Simon Kirby (2007) has located the pivotal point at which language meets culture. He notes Maynard Smith and Szathmáry's (1997) argument that language enables a system of cultural transmission, but he points out that this argument does not go far enough. Language does not need to be taught explicitly to infants: it "can be reliably acquired purely through the observation of instances of its use" (Kirby 2007: 10). Crucially, Kirby draws out the implication of this fact: language transmits not only semantic information, but also "*information about its own construction.*" Natural language is, in this respect, beautifully analogous to DNA, for it provides – to receivers with the innate disposition to receive it – its own means of replication and interpretation.

Cultural evolution requires an adequate explanation of how a representational system that was complex enough to carry a uniquely broad range of cultural information could have come to be shared by all humans. When we generalize over recurrent linguistic patterns, the result is that we discretize the language to which we are exposed, in ways that reflect its existing compositional structure. We produce, in other words, a system for representing information in discrete, hierarchically combined packages. When other people hear us using the language that we have learned in this way, they go through the same learning process. In this way, a biologically prepared species can acquire a complex, culturally evolved system of representational communication: once natural language has been acquired, it creates receivers with the capacity to interpret and implement the information that it carries. Language acquisition, in supporting the exchange of discrete, compositional cultural representations, provides the substrate for cultural evolution.

6

Language, Thought and Culture

What effect does language acquisition have on human cognition? Natural language evolved primarily for communication, but it is not only a conduit for information: it also shapes the ways in which we think about the culture that it enables us to acquire. We can receive information from a language only if we discretize that information in the same way that the language does, and this means that each language shapes the ways in which we interpret and respond to its content. Conversely, it does not enable us to interpret information that is represented in a different system. This chapter explores how, in facilitating the transmission of the local culture, local languages restrict their users to the informational content of that culture, as well as to a particular way of thinking about it. It also asks what happens when local language users come into contact with speakers of a different language, and it finds that there may be an impact on both the speakers and the languages involved. In these ways, we can begin to see the emergence of novel languages and different cultural frameworks, which prepared the way for the artefactual languages that were to follow.

Language and Cognition: Emotions

We have seen how, under primarily communicative pressure, natural language has evolved to be innately learnable by human children, who in turn have brains adapted to its innate acquisition. We have also seen that human cognition must discretize information in order to receive it: we cannot respond to a continuously varying input unless we form discrete representations of its varying states. One example of this process can be observed in toddlers' inability to cope with their own continuously varying emotions, and the way in which their parents can help by providing them with an emotional language (Distin 2006b: 101–3). A variety of research (for a review, see Lieberman et al. 2007: 421) indicates that giving names to feelings – a process known as affect labelling – helps to alleviate emotional distress. Functional magnetic resonance imaging has been used to confirm this fact and to investigate the neurocognitive pathways for the process (Lieberman et al. 2007). From a representational perspective, we can see that when adults give linguistic labels to both their own feelings and the emotions that they observe in their child, they teach her how to discretize the continuously varying range of her own and others' emotions. In this way, they can help her to move from a position in which she does not know how to respond to the continuously varying input of her feelings, to one in which she can cognitively process her emotions using the linguistic labels that she has been given.

As Chomsky (1987: 420) has pointed out, capacity and constraint are two sides of the same coin: the innate ability to do something in one way is, if your glass is half empty, an innate inability to do it in any other way. When natural language acquisition opens our minds, enables us to absorb and manipulate the human culture that it carries to them, and gives us the capacity to think in new ways, conversely it also limits us to the ways that it introduces. At a conceptual level, "the languages we inherit, how we represent and express things, package the world in a certain way. Just as with cartoons of political figures, the distinctions encoded in language call forth a certain noticing; they call attention to some things and not others in the environment within and around one" (Price and Shaw 1998: 189).

We should beware following this track to a wholly Whorfian terminus, however. Although there is plenty of evidence, of which this chapter will sketch a little, that language learning moulds our cognitive structures, it is also clear that the basic shape of the construction material has been provided by our genes. Research indicates, for example, that the semantic structures of emotional vocabulary are broadly shared across different natural languages (Romney, Moore and Rusch 1997), and this indicates in turn that there may well be a universal range of human emotions, which natural languages consequently discretize in very similar ways. Similarly, there is evidence that the colour terms across different languages tend to cluster around certain universal points on the spectrum, and this suggests that there is a universal cognitive basis for colour language and colour memory (Regier and Kay 2006).

Language and Cognition: Colour Perception

Nevertheless, within the genetic constraints that nature provides, there is a great deal of evidence that nurture, in the form of language learning, can have significant effects on our cognition. The reason our primary language so shapes our cultural assumptions is that linguistic structures discretize not only the media in which they are realised but also the information that they carry. Languages are representational systems, which discretize information in a particular way, and we can acquire information from a spoken language only if we learn to discretize sounds in the same way that it does. Each language brings information to us in chunks, which we can acquire only if we break them off at the same points that the language does.

This conclusion is supported by evidence from categorical colour perception. Although there are universal foci for the colour terms that are used in every language, different natural languages have different numbers of colour terms and draw the boundaries between colours in different places. Research has found that "linguistic differences actually seem to cause, rather than merely correlate with, cognitive differences" in how speakers of different languages apprehend colour (Regier and Kay 2006). In other words, the ways in which speakers discretize the spectrum are partly shaped by the ways in which their natural language discretizes the spectrum: "it appears that it can be said that nature proposes and nurture disposes" (Regier and Kay 2006).

Language and Cognition: Language Learning

A similar phenomenon can be observed at the most basic level of language learning. As Stephen Levinson (2003: 33) says, "it seems fairly self-evident that the language one happens to speak affords, or conversely makes less accessible, certain complex concepts." Levinson (2003: 43) argues that vocabulary helps us to think by recoding the multifarious content of our thoughts into larger chunks, "as a method of increasing computational power by getting around the bottleneck of short-term memory." He points out that both the form and the content of human languages vary profoundly (2003:29): different languages recode perceptual and conceptual input in very different ways from one another, and semantic differences of this kind are bound to produce corresponding cognitive differences (2003: 41–2). The semantics of the languages that adults speak turn out to be tightly consistent with the ways in which they think (2003: 42–3).

In fact, the process of learning our primary language will alter forever the ways in which we hear or attempt to speak other natural languages, impeding our ability to recognise their rhythmic patterns or to pronounce their distinctive sounds. As a result of exposure to the rhythmic characteristics of their native language in utero, newborn babies are able to distinguish between languages that do and do not have that rhythmic pattern, and by five months, they are additionally able to distinguish their native language from others that share the same rhythmic pattern (Bortfeld 2002). Another early change occurs in infants' categorical perception of speech sounds. At birth, they are able sort the speech sounds of any language into separate categories, but by six months, as we saw in the previous chapter, infants' speech perception has already been altered by exposure to the local language (Kuhl and Meltzoff 1997: 15) and they have begun to lose the ability to discriminate some of the sounds in foreign languages. Adults are unable to distinguish phonetic contrasts in any language other than their own (Kuhl and Meltzoff 1997: 9).

Language and Cognition: Spatial Cognition

There is evidence that the culture-specific language that we speak can even "play a significant role in structuring, or restructuring, a domain as fundamental as spatial cognition" (Majid et al. 2004: 108). Whenever we need a specified direction for one object with respect to another, we use a culturally variable coordinate system to code the objects' spatial locations. Different languages use, with varying frequencies and application ranges, three separate frames of reference. Relative frames of reference are egocentric systems in which coordinates are matched to our own orientation: we might say, for example, that the pen is to the right of the book. When using an intrinsic frame of reference, in contrast, objects are parsed into their own major parts, and coordinates are given relative to the named parts of a certain landmark object: the pen, in this case, might be described as being "beside" the book rather than to [our] right. Finally, absolute frames of reference use fixed bearings, like compass points, so that the pen might be described as being to the south of the book. In small-scale, "table-top space," English speakers use only the first two, and reserve absolute frames of reference for larger-scale, geographical descriptions (Majid et al. 2004: 108). Australian speakers of Guugu Yimithirr, however, use only absolute descriptions: neither a relative nor an intrinsic frame of reference is available in their language – not even for body parts (Majid et al. 2004: 108). Although this seems alien to Westerners, for whom the relative frame of reference feels most natural, the authors show that there is no evidence to suggest that the relative frame of reference is privileged in child development: on the contrary, all three frames of reference are acquired by children with comparable ease (2004: 112–13).

Once children have learned their local language's preferred frames of reference and their range of applications, experiments demonstrate that their use of nonlinguistic frames of reference, and of pointing gestures, will align with their linguistic preferences (2004: 110–11). Majid and colleagues (2004: 113) conclude that the categories of spatial cognition are variable and that "linguistic diversity aligns with cognitive diversity, as shown in people's language-independent solutions to spatial tasks and unselfconscious gestures accompanying speech," adding that their work "therefore contributes to the emerging view that language can play a central role in the restructuring of human cognition."

Language and Culture

Further support for the role of language in shaping cognition can be found in a recent body of psychological research into the effects of language learning on cultural values and behaviour. It seems that language acquisition provides learners with access to a particular culture, and that their subsequent use of that language affects their behaviour by priming them to interpret the world through a particular cultural lens (Hong et al. 2000; Benet-Martínez 2006).

These authors' conclusions are based on a dynamic-constructivist approach to culture. This approach, in keeping with a cultural evolutionary view, assumes that culture is not an integrated structure which forms an overall worldview, but is rather an associative network of discrete "knowledge structures" (Hong et al. 2000: 710), which affect how an individual will interpret the social world (Benet-Martínez 2006). Individuals can become multicultural, and even acquire mutually contradictory cultural elements, because novel cultural elements neither replace existing ones nor are blended with them. Instead, acquired aspects of culture influence behaviour to the extent that they are cognitively accessible and relevant to the situation.

There is abundant evidence from cognitive and social psychology that mental constructs are accessible to the extent that they have been activated by recent use. Researchers have tested this hypothesis using experiments in which they manipulate participants' exposure to a prime, which is a word or image that is related to a particular construct, and then measure the extent to which participants' subsequent interpretations of a different stimulus have been influenced by the prime (Higgins [1996] gives an overview of this research). The conclusion is that an individual may acquire many different units of information, some of which might even contradict one another, but the ones that will be most likely to be put into effect are the ones to which his attention is drawn by external stimuli, or primes.

The application of this evidence to the process of cultural frame shifting is clear. An immigrant from an entirely different culture will begin to think and behave as if she were a member of the host culture only if the host culture's interpretative frames are chronically accessible to her (Hong et al. 2000: 718). If she should choose to surround herself with stimuli that prime her native culture, however, then even if she acquires some of the cultural constructs of the host culture they are unlikely to have much influence on her thoughts or behaviour.

In particular, Hong and colleagues (2000: 717) view language as "an effective means of activating cultural constructs." There is a substantial body of evidence in favour of the view that language affects bilingual individuals' psychological responses, in areas such as such as personality measures, values, self-concept, emotional expression and other-person descriptions. For instance, there is evidence-based support for the anecdotal claim that "bilinguals express different personalities when they speak in different languages" (Ramírez-Esparza et al. 2006: 100). It appears, for example, that bilinguals' responses to value-related surveys are influenced by the language in which they are answering. One possible explanation is that when they change between languages, bilingual individuals undergo a cultural frame switch. This explanation is viable because bilinguals also tend to be "bicultural" (2006: 100), as each language is associated with a different cultural system (Hong et al. 2000: 717).

From a dynamic-constructivist perspective, empirical evidence has emerged that the acquisition and use of a specific natural language both enables us to learn a particular body of culture and restricts us to seeing the world in that particular way. Multicultural individuals have access to more than one cultural meaning system, and they shift between cultural frames in response to ambient cultural cues (Hong et al. 2000: 710), of which language is among the more significant. The use of a particular language influences a speaker's behaviour by priming him to interpret the world in one of the ways that he has acquired.

Contact Linguistics

The previous section has shown how encounters with novel languages can have an impact on the cultural frameworks and even on the psychological traits of the individuals involved. What effect, if any, will these encounters have on the languages that are brought into contact in this way? There has been considerable research into the impact of one learned language on another. "Whenever people speaking different languages come into contact, there is a natural tendency for them to seek ways of bypassing the communicative barriers facing them by seeking compromise between their forms of speech" (Winford 2003: 2). The outcome of such contact might be anything from insignificant lexical borrowings to the creation of a new language, depending on a combination of linguistic and social factors.

The primary mechanisms by which languages are reshaped when they come into contact with one another are known as borrowing and imposition (Van Coetsem 2000). Any language can be either the source or the recipient of borrowings and impositions, and the nature of the transfer that takes place will depend on which language is dominant for the speaker (Winford 2005: 55).

Linguistic borrowing takes place via the agency of speakers for whom the recipient language is dominant, as when native English speakers use phrases like "cul-de-sac" or "a priori" when speaking English. There is a tendency for speakers to "preserve the more stable components of the language in which they are more proficient" (Winford 2005: 7), and on the whole the stabler components comprise structural features like phonology and grammar, whereas vocabulary is less resistant to change. For this reason, when a person is speaking his dominant language, the material that he borrows from another language consists mostly of imitated vocabulary. A speaker who sticks to the recipient language's structures needs to know only as much of the source language as he wishes to borrow – and this could be as little as one word. These borrowings run the risk of mutation, as when a Japanese speaker uses the English-derived word *sangurasu* for sunglasses (Winford 2005: 8), but in essence they are lifted intact from the source language, retaining the meaning that they had there and having no structural impact on the recipient language.

Linguistic imposition, in contrast, takes place via the agency of speakers for whom the source language is dominant, as when native German speakers use a phrase like "I would suggest him to go" (Winford 2005: 10) when speaking English. Because speakers tend to preserve the linguistic structures of their dominant language, imposition tends to affect the grammar of the recipient language rather than just its vocabulary (Winford 2005: 7). Material from the nondominant recipient language is adapted to the rules of the source language, which is dominant for the speaker (Winford 2005: 11). Consequently, imposition relies on a great deal more second-language learning than does borrowing. A speaker who imposes source-language structures on a recipient language needs to have learned a great deal of the recipient language's vocabulary.

When it is less clear which of a bilingual's languages is dominant, both mechanisms of agentivity will be at work. Winford (2005: 34–42) argues that mixed languages like the Turkish-influenced dialects of Cappadocian Greek have arisen from a combination of recipient-language and source-language agentivity, incorporating both borrowings and impositions at many linguistic levels. Such a degree of recombination could only be supported by speakers who are able to make sense of both languages, and indeed the Cappadocian region "was characterized by a high degree of bilingualism" (Winford 2005: 41). Winford (2005: 35–40) produces extensive evidence that Cappadocian Greek has some linguistic features that can best be explained as imposition by Turkish-dominant bilinguals and others that have been imposed by Greek-dominant bilinguals; indeed, the same person might, at different points in his life, be counted as either.

Metarepresentation

This chapter has explored some of the ways in which natural language restricts its users to a particular way of thinking about the information that it enables them to acquire – but it has also revealed how contacts between speakers of different natural languages can produce cultural and linguistic change. This raises the possibility that we can begin to escape the cognitive and psychological restrictions of our native natural language.

It is revealing, in this context, to reflect on the awareness that we gain, when we become competent in a second natural language, of the influence that our primary language has had on our ways of thinking. The process of learning a second language increases our metalinguistic awareness, drawing our attention to features of our primary language that previously we were not even able to see, and increasing the efficiency of future language-learning strategies (Jessner 1999). Comparison between two alternatives helps us to see the key features of each, and when we make these sorts of comparisons the subject of our thoughts – the content of our representations – is not the information that the symbols represent, but the symbols themselves. We are *meta* representing: thinking about the ways in which we are representing information.

The content of a representation is, to put it clumsily, the thing that it is a representation *of*: an object in the world, perhaps, or an abstract concept or an action. A metarepresentation is a representation of another representation. Its content is that other representation, and crucially this includes information about both form and content. The ability to metarepresent is the ability to recognise the distinction between the two: to reflect on the connection between a representation and the information that it represents. The information that evolves, when we metarepresent, is information about *how we represent*. To put this another way, once we start comparing the representational features of different languages, the two systems effectively begin to compete with each other, under representational pressure.

As the briefest glance at modern culture makes clear, our cognitive escape route from the restrictions of our native language has not been restricted to other natural languages. Limited as it is by the length of the critical period and by the human capacity for learning, natural language has become, over time, inadequate to the representational task that it was originally set. If language is to account for cultural evolution then we need to look beyond natural language to the artefactual languages that have evolved in its wake.

PART III

THE INHERITANCE OF
CULTURAL INFORMATION

Artefactual Language

7

How Did Artefactual Language Evolve?

The evolution of natural language provided humans with a shared representational system, enabling them to acquire information from anyone who used that system. It was never a perfect representational system, because its evolutionary impetus had been provided by cooperative communication, which leaves room for the immediate checking and correction of intended meaning, without relying on absolute precision in the message received. Nor is it a limitless system: its capacity is restricted by users' cognitive abilities as well as by the length of time available to them for learning the system. Yet the result of natural language evolution was, despite the system's limitations, an explosion in the amount of information that early humans were able to trade. And although a lot of human culture does not depend on anything other than verbal and imitative abilities to be passed on to the next generation, there is only so much information that we can hold and manage in our brains alone; even in our collective brains. Eventually there came a point at which there was too much cultural information available for individuals to manage reliably using memory and speech alone. In this chapter and the next I explore how, when cultural information had expanded beyond the point that the brain could manage independently, humans began to develop artefactual symbols in order to represent – that is, to store and manipulate – the excess cultural information.

My claim is that, as the quantity of shared information increased, a new selective pressure emerged. Whereas natural language had originally evolved under the *biological* pressure for a cooperative species to be able to *communicate* more effectively with one another, now there was *cultural* pressure for the shared language to be able to *represent* information more effectively. As groups cooperated on increasingly complex cultural projects, there may sometimes have been a biologically adaptive advantage to members who had the ability to represent, in more persistent and unambiguous media than speech, information about plans, methodology and the key elements of the end product. Nevertheless, unlike natural language, which evolved for the immediate expression of our thoughts to other people, this chapter will show that artefactual systems of representation evolved under what we might call representational pressures, and that these pressures were cultural rather than biological.

In addition, this chapter presents an extended discussion of the evolution of writing. Writing, which is an artefactual representation of the spoken language, is perhaps the iconic example of an artefactual language. The essence of this book's thesis is that culture's origins lie in natural and artefactual languages, each of which has coevolved with its semantic field and the medium in which it is represented, to produce (via human agency) the variety and complexity of human culture that we see today. I have already sketched an explanation of how cultural evolution, obeying the biological imperative to improved communication, might have produced natural language. Here I explain how cultural evolution, obeying the cultural imperative to improved representation, might have produced that other hallmark of our species, the written language.

The Biology of Artefactual Representation

Natural language emerged on the back of a range of preadaptations, including the capacities for representation and voice control, and the urge to communicate with one another. The genetic hardware was ready, and natural language is something that humans learned to do as a result of those innate capabilities, rather than a novel genetic innovation on top of what was already in place. The cultural evolution of natural language enabled humans to make much more efficient use of their existing cooperative, representational and vocal capacities, and this had biologically adaptive advantages because it enhanced humans' ability to communicate.

At first glance, it might seem that the selective advantages to humans of increasing competence in cultural areas like technology and medicine, for example, might well have led to a similar degree of coevolution between the brain and artefactual languages. But for several reasons this coevolution has not gone so far as the link between the brain and natural language.

First, much of the coevolutionary work had already been done. Artefactual languages emerged on the back of existing mental and physiological capacities: symbolic and pattern-recognising abilities, increased sociality, clever hands and so on. Literacy skills, for example, are supported by preadapted brain structures rather than by structures that evolved to mediate reading and writing (Petersson, Ingvar and Reis 2008), and more generally, the same abilities that enabled us to learn natural language could also be put to work on the acquisition and development of artefactual languages. Archaeological evidence supports this conclusion. The oldest fossil evidence for anatomically modern humans is between one hundred thousand and two hundred thousand years old in Africa, but there are striking similarities between the archaeological records of these early modern humans and those of the Neanderthals (Mithen 2000: 210–11). There is evidence from this period of some concern with the dead (White 2003), for example, but "it is not until between 60,000 and 30,000 years ago that the archaeological record is transformed in a sufficiently dramatic fashion to indicate that a distinctively modern type of behaviour and mind had evolved" (Mithen 2000: 211). In other words, as Mithen emphasises, there was a space of at least seventy thousand years between the emergence of *Homo sapiens* and the appearance of modern behaviour and thought – a development that was not associated with any new change in brain size.

Secondly, artefactual language evolved at precisely the point when our brains' innate capacity to hold and manipulate information ran out of steam. If its function is to manage (and indeed facilitate the emergence of) the cultural information that we cannot manage with brains and speech alone, then its purpose is, in other words, precisely not to be innate. Whereas there is an adaptive value in learning the collective communication system before mating age, artefactual languages are most useful to physically mature humans, and indeed the acquisition of artefactual languages can go on throughout our adult lives. At a physiological level, just as children cannot learn to produce spoken languages until their vocal systems have developed sufficiently, so they cannot learn to produce written languages until their fine motor control has developed sufficiently. They can of course understand natural language before they can speak it; and similarly they read before they can write – but they speak before they can do either. Indeed, young children are at several disadvantages in the acquisition of artefactual languages. For instance, until they are able to develop the conceptual apparatus that an artefactual language helps to build, there is nothing for children to gain from learning how to represent the informational content that is supported by that apparatus.

Finally, no matter how useful the innate acquisition of artefactual languages *might* have been to humans, the cultural evolution that resulted from their emergence has been so swift that it has overtaken biological evolution, which can no longer keep up with it. Reading and writing are skills that emerged only six thousand years ago – a blink of the biological evolutionary eye – and further evidence for their reliance on cultural rather than biological evolution comes from the fact that, even today, one in five adults is illiterate (Petersson et al. 2007).

Against the background of an increasingly competent species of communicators, who were able to share a rapidly expanding amount of cultural information, it is easy to imagine that the new culture would have brought adaptive advantages, but we have seen that where there is advantage, there is not always adaptation. Culture is a good trick for humans, and its expansion was dependent on the emergence of ever-better methods of representing the information on which it is built. But this should not be taken to entail that those representational methods are innate. Artefactual language evolved culturally, on the back of both innate biological preadaptations and culturally evolved natural language: the driving force was representation, but this was primarily a cultural rather than a biological pressure. Artefactual language was certainly not in conflict with genetic interests, and indeed will have brought certain advantages, but it was primarily a cultural innovation. The apparently gratuitous complexity of artefactual languages enhances our capacity for representation, on which our culture is based, but the sorts of rules to which artefactual languages conform will not need to be innately learnable by human children. We have the luxury of learning them consciously, because we do not need them in order to communicate with one another, only to handle information.

The Power of Metarepresentation

How did natural-language-using humans begin to develop alternative, artefactual representational systems? The very first artefactual symbols were alternative means of representing information that had previously been carried by natural language. In mathematics, for instance, the scientific notation is preceded, both in language history and in individual learning patterns (Hurford 2001: 2), by natural language numeral systems like the traditional northern English *yan, tan, tethera* (one, two three) for counting sheep. Although numeral systems are explicitly taught, rather than acquired innately like the rest of natural language – which suggests that they exist on the very fringes of what our collective mentality can manage without external props – they still precede the written scientific notation. What this means is that with the creation of the very first artefactual symbols there emerged two alternative ways of representing the same information: the spoken words and the written symbols. As the alternative systems began to compete with each other, under representational pressure, each artefactual language began to evolve towards more efficient representation in its specific cultural area. Mathematical notation, for instance, is far more compact than the vernacular: compare the number 260 with the explicit representation of the addition operation and the names of the powers of the base number in the more cumbersome English phrase "two hundred and sixty" (Hurford 2001: 2). In culture, as in biology, what evolves is information – and when we started to think about how we were representing that information, we kick-started the evolution of artefactual languages.

It is when we move from natural to artefactual language use, changing our priority from communication to representation, that our metarepresentational capacity really begins to bear cultural fruit, for it is not until we acquire alternative means of representing information that we also learn new ways of thinking about it.

A language is a system for representing a certain portion of information in a certain medium, and its structures will be shaped by both that semantic field and its medium. Natural language, for instance, is designed to communicate human thoughts in the serial medium of human speech, and its structures (e.g., phonology, syntax) have coevolved with both that medium and those thoughts. Characteristics that make speech particularly well suited to the outward expression of inner thoughts include the facts that it can be swiftly produced and easily received; that there is a prepubescent period critical to its acquisition, which is facilitated by a raft of innate preadaptations; and so on. Its coevolution with the human brain and physiology makes it particularly well adapted to the serial communication of the content of human thoughts. The ways in which it discretizes information has, conversely, a massive impact on the nature of those thoughts.

Like natural languages, artefactual languages inevitably shape the ways in which we think about their cultural content. Each artefactual language evolves to represent a specific area of culture; and that area of culture evolves, in turn, to be conceptualised in ways that are representable by the relevant artefactual language. The evolution of artefactual languages vastly increases and enhances the ways in which we can think about the areas that they represent; but conversely they also limit us to those ways. It is the power of metarepresentation that enables us to escape these limitations. Humans are uniquely able to think not only about the content of our representations but also about the representations themselves. We can lift information from its original context and reflect on it in the abstract. We can learn not only new information but also new ways of representing it; new media in which to embody it.

This is significant because, as Robert Aunger (2002: 157) has emphasised, the medium in which information is stored is enormously significant for evolutionary dynamics (though note that Aunger would deny that information can be replicated across media). A poem might be printed in a paperback anthology, spoken aloud at a recital, preserved on a vinyl record or accessed on a Web page, and each version will have a different impact on the longevity and stability of the poem's preservation, the accuracy and fecundity of its replication, the potency of its emotional effects, the size of its potential audience and so on. The system in which information is represented will impact on a similar range of factors. The poem might be represented in the English spoken language, in Chinese Sign Language, in written Nynorsk or even in ASCII. All of these alternatives will, in combination with the medium in which it is realised, have a particular profile of effects on the evolutionary dynamics of the same piece of cultural information. Each language will "call forth a certain noticing" (Price and Shaw 1998: 189); each medium will deteriorate at its own rate. Some media, and some systems of representation, are inherently better than others at ensuring the long-term, faithful preservation of cultural information. Others carry more potential for swift and extensive transmission. Each endows the information that it carries with a particular evolutionary scope – intimately connected with which are its particular evolutionary limits.

The significance of metarepresentation is that it frees cultural information from the limitations of any one medium or language. Many of our concepts depend for their coherence on the language in which we originally encounter them: we cannot think about zero, for example, if we're stuck with representing mathematical information in Roman numerals. To look at this from the opposite angle, it's clear that when we learn a new language our minds are opened to a whole new swathe of information. The ability to lift information from its original context; to transfer it between languages and media; to acquire new languages; and then, when you have opened the doors to the information that they bring with them, to lift it out of that original context into one with different scope, different limitations – it is this metarepresentational ability that is the driving power behind cultural evolution. It is this that enables us not only to acquire information but also to think about which information we have acquired: to recognise and escape the impact of the information currently underlying our behaviour. And it is this, too, which provides the mechanism for the evolution of cultural languages.

The Evolution of Writing

Artefactual languages, like natural language, evolved and continue to evolve iteratively. As a range of artefactual symbols increases in scope and use, so its structure will develop as users interpret and internalize the rules governing the ways in which it is currently being used, and as a bottleneck of transmission develops between those who know the symbols and those who don't. The shared use of an artefactual language depends, just as the shared use of natural language does, on the human instinct for cooperation: a willingness by all participants to abide by shared rules. Some artefactual systems, like present-day traffic lights, are so limited that we simply learn each unit holistically: they are noncompositional in structure. If, however, the range of symbols is large enough that only a small subset is observed by new learners, then a compositional structure is needed to ensure that the system is learnable and expressive. The archaeological history of writing provides ample evidence that artefactual languages evolved culturally under selective pressure for better representation.

Proto-Writing and Social Context

There is evidence from around thirty thousand years ago that our ancestors were expressing and recording elements of their culture in cave paintings and rock carvings (Lewis-Williams 2003), and the earliest known mathematical artefact, a tally or calendar stick known as the Lebombo bone, dates from around the same time (Bogoshi, Naidoo and Webb 1987). Later, around ten thousand years ago, there is evidence of prehistoric administrative artefacts like clay tokens and stamp seals (Schmandt-Besserat 1996: 7). Finally, from the end of the fourth millennium BC, we can see a move from these preliterate means of representation to systems of proto-writing and subsequently to the written representation of oral language (Schmandt-Besserat 1996: 1).

It is often assumed that the function of writing is to represent a particular oral language, but in fact "the influence of the structures of language on a system of writing becomes weaker the further one goes back in its history" (Damerow 1999: 2). There are significant differences between systems of proto-writing, such as the proto-cuneiform script found in ancient Babylonia, and oral systems of representation. For example, these scripts do not code information phonetically: where combinations of signs are used, there is frequently no relation between the sign combinations and that which the combination depicts (Damerow 1999: 9–10). Another difference from the spoken language, which is a series channel of communication, is the largely hierarchical organization of proto-cuneiform texts (Damerow 1999: 7), which is a consequence of the fact that the information they represent is mostly economic. This restricted semantic field is another way in which early writing systems differ from natural language (Damerow 1999: 8–9): the original context of application of a system of proto-writing seems to have had a strong influence on the syntax and semantics of the writing system that subsequently developed (Damerow 1999: 13).

Indeed, if writing can usefully be seen as an artefactual language that emerged under representational pressure, then we should expect that its structure would reflect its content, for cultural languages are systems in which a certain portion of information (the semantic field) can be represented in a given medium. The content of proto-writing was mostly economic, and its structure reflects this fact. The revealed importance of the context of application, in shaping the structure of the earliest written system, exactly parallels my emphasis on the importance of informational content in shaping the structure of any artefactual language.

From Proto-Writing to Writing

The archaeological evidence about the transition from proto-writing to writing provides further support for this view of writing as an artefactual language that culturally evolved under representational pressure. When investigating Near Eastern clay objects from eight thousand to ten thousand years ago, Denise Schmandt-Besserat (2002: 6) discovered an increasingly complex system of tokens, which represented, in one-to-one correspondence, various goods. These tokens were first used around 8000 BC, at the time when the advent of agriculture brought with it the need to keep track of goods (Schmandt-Besserat 2002: 6); earlier hunter-gatherer societies had tallied time's passing but had no need to keep track of goods, which were not accumulated (Schmandt-Besserat 1996: 103). Later, around 4000 BC, manufactured goods were tracked in an urban context using more complex tokens with surface markings. There was no link, at this stage, between trade and the use of tokens. Rather, tokens were used to keep records in the context of a newly ranked society in which a complex redistributive economy had developed: complex tokens "were developed to handle with greater efficiency and precision the larger volume of goods generated by taxation and the levy of tribute" (Schmandt-Besserat 1996: 110).

The only sensible explanation of why the use of these tokens should have become so widespread is that they had representational advantages over human memories. Accounting information has a selective cultural advantage if it is represented in collections of clay tokens, because it is preserved more accurately and persistently than the human brain can manage. The tokens were a long-term repository for information that the farmers would otherwise have to track in their memories, and they represented this information in a shared system that enabled anyone, once they had grasped the system, to understand what any given collection of tokens meant. "It thus became possible to store with precision unlimited quantities of information concerning an infinite number of goods without the risk of depending on human memory" (Schmandt-Besserat 1996: 93–4).

Information represented in the tokens also had a selective advantage over information represented on tally sticks or bones, because the tokens had a greater semantic capacity than the sticks, a rule-based organisation and recognised (arbitrary) links between form and meaning. Whereas the marks on tally sticks were meaningless to anyone but their maker, the token system made use of different shapes, each of which had its own, known meaning. Clay was abundantly available, and sufficiently plastic and easily worked when wet that it enabled unskilled individuals to create recognisable shapes, even without tools – shapes that, furthermore, became permanent once the clay hardened.

The early token system, however, was cumbersome and noncompositional, and it relied on external methods of preserving token collections. Fourth-millennium accountants began to seal tokens in hollow clay envelopes, which were sometimes marked with the impressions of the tokens that they contained. Another change that took place rather suddenly at this time was that the number of different types of clay tokens increased dramatically, from about a dozen to more than three hundred. Why should the farmers have gone to the trouble of enclosing tokens in a clay envelope? The advantage of the new system was that it provided an "orderly and tamper-proof" accounting record of collections of tokens, which represented objects in a one-to-one correspondence (Schmandt-Besserat 2002: 7). Representations of accounting records thus became more permanently accurate and more universally trusted. When envelopes were marked, before being sealed, with impressions of the tokens that they contained, this enabled officials to keep track of an envelope's contents without breaking it, which made the economic representations even more permanent. Information about collections of tokens is preserved with longer-lasting accuracy, when it is represented in marks on sealed clay envelopes, than either the human brain or the unsealed collections can manage. It gains an additional selective advantage from this longer-lasting accuracy, which is that it is trusted more by its human users than the unsealed token collections are, and this gives the sealed collections an advantage in the competition for human attention.

This transition, from tokens to envelopes, also provides historical evidence of how the nature of the medium in which a representation is made coevolves with the nature of the representational system that is used. Instead of a three-dimensional representation of an object, as the clay tokens had been, the envelopes were impressed with two-dimensional representations of the representational tokens that they contained. The mere fact that information was being represented in two dimensions on an envelope, instead of in three-dimensional tokens, had given rise to a new, metarepresentational system: a system in which information was represented about collections of representational tokens. It did not take long for the new system, in its turn, to affect the medium in which it was realized. Within a couple of hundred years, the envelopes' contents were becoming obsolete: marks on a solid clay tablet could serve the same purpose as marks on a hollow envelope with tokens inside; and information that is represented on clay tablets has a selective cultural advantage over information that is represented on clay envelopes, because the tablets are easier to produce and less fragile than the envelopes.

Yet this transition in medium, from envelopes to tablets, itself transformed the representational system once more: whereas the marks on the envelopes had represented the tokens that they contained, the marks on the tablets represented the items that the tokens had once represented. "Thus it took no fewer than four inventions – tokens, envelopes, markings and tablets – and about 4,000 years to fully reduce three-dimensional tokens to written signs" (Schmandt-Besserat 2002: 7).

Like the marks on the old envelopes, however, the marks on the new tablets were still being impressed by tokens that stood in a one-to-one correspondence with the items they represented. What happened next is the clearest indication that archaeological evidence can provide of the representational pressures to which the writing system was subject – and it all happened so quickly that it would be ludicrous to suggest that the consequent representational evolution was genetic in origin. Little more than a century later, the marks on the clay tablets were beginning to be made with a stylus rather than by impressions of the tokens. Information is represented more precisely in incised signs, made with a stylus, than in impressed signs made with tokens: although contextual clues were used to overcome the ambiguity that arose when the impressed token shapes were rather blurred (Schmandt-Besserat 1996: 68), they could not achieve the precision of an incised sign's representation of the profile and even the surface markings of a token.

Another selective cultural advantage of the incised marks over the token-impressed marks was that a greater variety of marks can be made with a stylus, so the new system had greater capacity as well as more efficiency. Indeed, at the same time as the marks began to be made with a stylus, they also lost their restrictive, one-to-one correspondence with the objects that they represented. A shorthand was developed for the representation of quantity, by dint of changes in the meaning of existing marks such as the signs for a small or large measure of grain (Peterson 2006), with the result that the sign for an object could be preceded by the sign for a numeral. "Here was a marvellous economy of signs: 33 jars of oil were expressed by seven signs ($10+10+10+1+1+1$ and 'oil') – rather than 33 signs" (Schmandt-Besserat 2002: 63). Another word for this economy is *compositionality*: with the increased amount of information to be represented, the written symbols now provided an effective bottleneck in its transmission from one person to another, and compositionality was the solution. Once information is represented compositionally, it has an even greater selective advantage because it is represented more efficiently, which makes it less vulnerable to copying errors and hence more attractive to human users.

Here, too, we can see the impact of the medium on the representational system; the impact of the representational system on the information that it can carry; and the impact of the information on the way in which it is represented. Because they were now made with a stylus, the marks could deviate from the old system of mimicking the shapes of the tokens with which they were impressed, and as soon as a range of different marks was available, different information could be represented: information about the people who had given or received the goods, for example, as well as about the goods themselves. The crucial step came in about 3100 BC, when the accountants of Uruk, in modern Iraq, invented abstract numerals, separating stylus-incised pictographs, which depicted commodities, from token-impressed numerals, which depicted abstract numbers. Previously, the signs on clay tablets, like the clay tokens that preceded them, had represented "concrete" numbers. These concrete numbers fused together the concepts of numbers and the particular items being counted, so that different items were counted using different tokens or signs, just as in some languages there are different number words for counting different items (compare the English words *trio*, *triplet* and *trilogy*): there had been no tokens or signs for the representation of abstract numbers. With this innovation in informational content, the nature of the system changed, too: with the introduction of abstract numerals, "the signs for goods and the signs for numbers could evolve in separate ways. Writing and counting generated different sign systems" (Schmandt-Besserat 2002: 63).

It is clear from this brief summary that information can gain a selective cultural advantage by being represented in one system rather than another. This representational selective pressure drives the evolution of representational systems as well as of the information itself. It also drives the evolution of the media in which these systems are realized. For example, "It is interesting that many techniques were devised for showing the token contents of envelopes" (Schmandt-Besserat 1996: 54): tokens were not only used to stamp the soft clay but were also sometimes attached to the surface of the envelopes; marks were made, alternatively, by a stylus, by a thumb, or by scratching the clay once it had hardened. Of all these variations, only token impressions and stylus markings survived, and it is easy to see the representational advantages that these held over, say, thumb marks. It is also interesting to note that although other methods of keeping token collections intact were used as well as envelopes, such as stringing them together, certain sorts of tokens were most often contained in envelopes, and consequently these were the tokens most often used to mark the envelopes. As Schmandt-Besserat (1996: 68) concludes, "This suggests that the fashion in which tokens were kept in archives determined the resulting script."

It is here that Schmandt-Besserat's work meets Damerow's emphasis on the social context of proto-writing, which is crucial for understanding how the more sophisticated artefactual systems of representation emerged to underpin cultural evolution. Damerow points out that the reason the early structure of proto-writing did not mirror the syntax, or encompass the vast semantic sweep, of oral languages, was that these early artefactual languages were not developed for the representation of speech. Rather, they emerged as a means of representing information in a particular area: administration and bookkeeping in the case of proto-cuneiform. By restricting the semantic field of a system of proto-writing, this social context shaped the structural and semantic development of the system of writing that subsequently emerged in each area. I would add that in this way, over time, there emerged a whole range of artefactual systems of representation, each one specifically tailored to the storage and manipulation of information about a particular cultural area, in a particular medium.

The most universal of these systems is, of course, writing itself. In her later work, which examines the interface between writing and art, Schmandt-Besserat (2006) focuses on the funerary, votive and dedicatory function of writing as it emerged from proto-writing. She emphasises the importance of the *name* in the Sumer region of Babylonia in the third millennium BC: the belief was that a person lived on for as long as his name was spoken. Supported by archaeological evidence from the Royal Cemetery of Ur, Schmandt-Besserat shows how funerary inscriptions were used as a means of permanently reciting names to the gods. She argues that this desire to be heard by the gods motivated the Mesopotamians to move on from inscriptions of a single name – which were nonetheless highly significant for their innovative use of phonetics – to simple sentences with subjects, verbs and complements. In this way they developed a "phonetic and syntactic system of communication able to reproduce speech" (2006: 5).

New media facilitate the emergence of new representational systems, which can carry different content, and so the interplay continues. At every step of the archaeological evidence we can see the representational advantages that the newly popular representational system or medium gave the information that it preserved, as well as the coevolution of informational content, representational system and medium.

Writing: Communication and Representation

Artefactual languages are obviously not only representational; they also play a communicative role, as indicated by the fact that they are shared systems of representation rather than idiosyncratic conventions like the personal shorthands that individuals develop in their private notes or diaries. Although artefactual languages have evolved under a representational selective pressure, the tension between their two roles can be seen, for example, in the metarepresentational function of writing. Just as impressed marks on the old clay envelopes were two-dimensional representations of three-dimensional representations, so writing is the artefactual representation of spoken representations of meaning. Metarepresentations of this kind have all the (dis)advantages of the system that they represent. They are semantically restricted, for example, to the informational content of the represented system: token impressions could only represent the accounting information that the tokens themselves had evolved to represent; it was not until the medium changed, from token on clay to stylus on clay, that the system was freed from the restrictions of the token shapes and could carry different information. Conversely, the *power* of the written word lies in the fact that it represents a system that has evolved to represent any thought that a human can have. And it is not only the breadth of its content that marks out writing as a particularly powerful form of representation. Because of its metarepresentational role as the representation of the spoken language, writing can be used (within the limitations of its writing-specific media) to communicate with other people over a distance or when speech is not possible.

There are times when our primary use of the written language is as a tool of communication: when we write a letter to a friend, for instance, or when we send a text message or an email. The communicative imperative in these cases is manifest in increased levels of ambiguity, as shown most clearly in the emergence of text-messaging language ("hpE bday! c u 2nite"). The immediacy of feedback via texting, emailing, instant messaging, social networking or blogging makes such processes particularly well suited to social communication.

Another communicative use of the written language is to annotate other artefactual languages, as, for example, when we add explanatory notes to a mathematical calculation ("divide both sides of the equation by $5y$"). Often, in such cases, we are using the written language to communicate ideas that a purely representational language cannot. Unsurprisingly, the efficiency of artefactual languages for representing information in a particular cultural arena is not matched by their capacity to manipulate information about anything else, including themselves or other systems of representation. We resort to natural language when we want to talk about artefactual languages or about comparisons between them and natural language, or to express our thoughts about how we are representing information or about how we are thinking. Natural language has evolved for just this communicative purpose, and one of the roles of writing is to transcribe what we say: once we are competently literate, we can use a written as well as a spoken code for communication with one another.

Nevertheless, the written language is not only a transcription of our spoken language. As well as the power of the spoken language, writing offers the representational advantages of its own media and representational systems: greater precision and accuracy than is normal in everyday speech, greater persistence, greater capacity, physical detachment from the humans who produce it, and so on. For these reasons, we can use writing to represent our thoughts, to work them out for ourselves and to set them out more clearly for others, as well as for the permanent record of our words. Writing's dual role can be seen in the sometimes-conflicting purposes of its punctuation, which not only illuminates the grammar but also guides readers through the sense and sound of what is written (Truss 2003: 68–82). This dual role can also be experienced by writers who reread what they have written and realise that it does not express their thoughts as unambiguously as they had intended. Then they must try to put themselves in their readers' shoes and represent their ideas in a form that can be universally understood. In other words, they must shift the aim of their writing from expressive to interpretative success: the purpose is not to get information across swiftly and accurately enough, but to get it across permanently and accurately. The written language is able to perform both functions. Just as other artefactual languages have evolved for the representation and manipulation of concepts that could not be managed so efficiently by natural language, so the written language serves our representational purposes. It has the potential to preserve our ideas in a permanent, unambiguous format, and in the evolution of jargon we can see the same sorts of conceptual tools as are provided by nonlinguistic symbols. Writing is a means of representing our thoughts as well as of communicating them.

Like all representational systems, the acquisition of literacy opens us up to new concepts and ways of thinking. When children learn to read, it both alters their phonological representations of the spoken language (Petersson et al. 2007: 791) and increases their verbal working-memory capacity (2007: 797). As long ago as 1904, Ernest Weber suggested that the acquisition of literacy might increase the language dominance of the left hemisphere (Petersson et al. 2007: 792), and more recent studies by Petersson and colleagues (2007) have confirmed that literacy affects the functional balance between the right and left inferior parietal region, such that literate individuals are relatively left-lateralized when compared with their matched illiterate controls. The influence of literacy on brain structure helps to explain the reciprocal and rather complex history of the relationship between spoken and written languages. Although the spoken language predates writing, the emergence of a written system has had a powerful influence on the structure, vocabulary and use of language as it is spoken. This is only to be expected when we realise that once we have learned a written version of our language, it provides an alternative with which we can compare our original, oral version. The resulting enhanced metalinguistic awareness leads to development in our use of both the spoken and the written version. As we saw in the previous chapter, contact between two languages can help to reshape either or both of them, depending on the extent of the receivers' familiarity with each.

Speech: Communication and Representation

It is also the case, of course, that the spoken word plays a dual role as a medium of communication and of representation: it might not carry out its representational role so effectively as the written word, but quite obviously it does have the capacity to carry information from one person to another. The cultural significance of this dual role, which the spoken as well as the written word plays, is best illustrated by the human proclivity for narrative.

Storytelling is, from a cultural evolutionary perspective, a good way of ensuring the retention and replication of information: we enjoy the process of hearing or reading a story, but we also expect to learn something from it. Stories are good ways of ensuring that information is learned, because they make links between their elements so that it is easier for us to move from one to the next. Experiments by Alex Mesoudi, Andrew Whiten and Robin Dunbar (2006: 407) demonstrate that linking events into a narrative produces better recall, and reveal that the most important factor in accuracy of transmission is the *social* content of information (2006: 413–14).

Surprisingly, their research also reveals that narrative drive, in the form of juicy bits of social gossip, is not an important factor for accuracy of information transmission; the mere fact that information is about social events is enough, and juiciness doesn't add anything to this factor. When presented with two stories that involved the same number of social interactions and social agents, participants showed no significant differences in either the quantity or the accuracy of their recall, even though one story was about a woman's journey to a swimming pool and the other was about her affair with her married college professor.

This can perhaps be explained, however, if we consider what might happen outside experimental conditions, when people are allowed to select what they talk about. From an evolutionary perspective, another function of stories is to grab our attention: they give us an incentive to listen to the message that they convey, because it is embedded in a format that we like. Stories provide the artistic spoonful of sugar that makes the informational medicine go down. On this view, the level of scandal involved in a piece of social information might not have any impact on the representational accuracy with which it is transmitted, but I would predict that in everyday life juiciness will have an impact on the frequency of transmission.

Storytelling thus illustrates the interplay between representation and communication. On the one hand, stories are a highly effective means of ensuring that the story teller or writer's message is communicated. A good story is in some ways the pinnacle of language use: it grabs the audience's attention, and cares more about the retention of a narrative pattern than it does about the reproduction of exact representational details. On the other hand, a good story does also play a representational role: the narrative pattern that it encourages us to remember is inseparable from the message that it is intended to convey.

Mesoudi and Whiten (2004) confirm the hierarchical bias in the transmission of cultural information: higher-level information has greater copying fidelity than lower-level details. They begin by presenting evidence from a series of experiments, over a period of more than seventy years, which "suggests that humans and some other species represent knowledge of routine events or stereotypical action sequences hierarchically, and tend to show better memory for, and imitation of, actions that are represented at a relatively high-level of that hierarchy" (2004: 6–7). Mesoudi and Whiten tested for this hierarchical bias in human cultural transmission, and their results confirmed that when events were described in terms of lower-level constituent actions and then passed along chains of participants, the lower-level descriptions were spontaneously transformed into descriptions from higher levels of the hierarchy. A lengthy description of a woman's trip to a supermarket, for example, which in its original form had included details about her parking, collecting a trolley, checking her list, looking for the fastest queue at the checkout, unloading items from the trolley onto the belt, and so on, was transformed by the fourth generation of recall into a description as brief as "Rachel went to the supermarket, got some food and went home" (2004: 22).

This hierarchical bias would explain how the narrative pattern of stories is preserved, as we hang on to their narrative structure while filtering redundant, lower-level details. Stories thus fit information for its ecological niche – which in this case is human psychology. We tell stories because narrative is a good way of organising, retaining and communicating information.

Storytelling is thus an important means of increasing our brains' capacity for, and accuracy in, the acquisition, retention and communication of the information that natural language transmits. As we have seen, however, the quantity and complexity of cultural information long ago outgrew the constraints of natural language. The capacity for metarepresentation, which had been an important factor in humans' innate ability to learn their native natural language, enabled them to continue learning new languages throughout their lives, including the emerging artefactual languages. There is archaeological evidence, as we have seen, that the evolution of writing was driven by the representational advantages of the newer media and representational systems over their predecessors. In this respect, writing is typical of all artefactual languages, each of which, as the next chapter will show, brings a variety of representational advantages to the cultural information that it transmits.

8

Artefactual Language, Representation
and Culture

For information to be shared between multiple receivers, all of the receivers need to understand the way in which its content is represented: they need to play a cooperative game in which they all obey the same linguistic rules. This is why the biological advantages of enhanced communication are so dependent on the cooperative tendencies of the communicating species. Given the preadaptations of our ancestors, it turned out that enhanced communication brought sufficient adaptive advantages to support the cultural evolution of communication systems with the complexity and scope of natural language. Natural language has a considerable capacity for representation as well as for communication, but unsurprisingly, it is best suited to the task for which it evolved: the immediate expression of our thoughts to other people via audible signals. For this reason, it has the representational disadvantages of transience, dependency on human memory, and semantic restriction to the cognitive structures that the human brain can support without external scaffolding. In contrast, the structure and media of artefactual languages offer a variety of representational advantages to the cultural information that they carry, just as DNA's double-stranded structure has greater stability than single-stranded RNA, and its complementary structure makes it particularly suitable for accurate replication and dispersal.[1]

Longevity and Fecundity

The most obvious representational advantage that artefacts have over speech is their persistence: they increase information's longevity. Any

---

[1] A comparison for which I am indebted to an anonymous reader for the Cambridge University Press.

representation lasts only as long as the medium in which it is realized. When I say something, if I have communicated successfully then its content has the potential to last for as long as the memories of anyone who heard it. When those people and I are dead, the things that I said will almost certainly have died with us. Although the oral tradition is notable for preserving poems, songs and stories from generation to generation, making the best possible use of the interaction between speech and memory, this is a very selective method of preservation, which excludes almost everything that most people say in their lifetimes. In contrast, although artefactual languages do exist whose primary media are as transient as speech – for instance, the positions of semaphore flags and the flashes or clicks of Morse code – most artefactual media offer information the culturally adaptive advantage of stability. As soon as I write down what I have to say, no matter how trivial or mistaken its content, it will last for as long as the medium in which it is preserved. When we see the history of artefactual media from the perspective of the information that they preserve, the representational pressure on their evolution becomes clear.

There are two ways to overcome the survival risks to the current generation of information: find ways of preserving it for longer or make more copies of it. In culture, the emerging artefactual media and the technologies for their production had massive implications for both. Marks chipped into stone can preserve information for thousands of years, but they take a relatively long time to make, and each person can only make one set of such marks at a time. Marks scribbled on a piece of paper are quicker to make, although paper does not last as long as stone. Once we interpose automated technology between the human information producer and artefactual representations of that information, we can increase the information's prolificacy to a level at which it effectively overcomes the limitations on its longevity. Previously, the processes of creating a recording medium such as papyrus or velum, and copying the information onto it, were so labour intensive that much recorded knowledge may have been written down only once (Hall 2006): if that copy was lost before anyone had transcribed it, then its content was also completely lost. The introduction of moveable type and the printing press, however, turned the production and replication of information into a massive industrial process, which insured written knowledge against easy loss.

At the same time as information was becoming easier to copy, it was also becoming easier to access. This was partly the effect of the new technologies: whereas we must scan linearly through scrolls, for example, books allow us readily to access any part of their content simply by turning the pages (Hall 2007). But this accessibility was also a result of the impact on literacy levels of the increasing availability of ever-cheaper documents. The coevolution of printing technology with increased levels of literacy enabled information to break through the one-to-few dissemination barrier: the early presses collated, preserved and disseminated existing knowledge to a previously unimaginable extent. Subsequent electronic processing technologies have further extended the dissemination and prolificacy of information, even enabling a computer's electronic circuitry to perform automated operations on its contents, with no subjective involvement from an individual human (Hall 2007).

But the loss or corruption of the last remaining copy of a piece of information is not the only risk, even for the most persistent of artefactual media. Rapid cultural evolution can discard a language or artefactual medium as easily as it can discard informational content. Information can be lost if the medium that preserves it outlives those who know how to interpret the language in which it is represented: there are no receivers for the information that is represented in an ancient system like the script on the Cascajal block, for example (Rodríguez Martínez et al. 2006). Information can be lost equally effectively by realization in obsolete media like 5.25-inch floppy disks, which have largely outlived the technology that is needed to extract the information that they preserve.

It is interesting to note that, as cultural content has expanded over time, the interaction between its longevity and fecundity has, to a certain extent, mirrored the interaction between these features in genetic evolution. In the early days of cultural evolution, when natural language was supported by only a few artefactual proto-languages, there were fewer copies of any given piece of cultural information, but a great deal of care was taken to ensure that each copy was preserved. This was a time when few people could understand and use the available artefactual systems of representation – so the resource of human attention was pretty scarce for information that was represented in artefactual languages – and the cultural environment was largely predictable. As literacy levels have increased, however, and technological innovations facilitate cultural fecundity, we can see that many more informational copies are made, but each individual copy is less valued and less well preserved than it would have been in the past. We invest less now in information's long-term preservation than we do in its immediate dissemination, partly because the cultural environment is changing so rapidly that much current information will stand little chance of success in the future.

Biologists will recognise in this brief sketch a pattern similar to the ways in which "survivorship and fecundity may be adapted to different aspects of environmental variability/stability" (Irwin 2001) in the natural world. As ever, there are limits to the extent of the analogy between culture and nature, and like genetic models of life-history patterns, we should not expect to be able to account for every facet of culture using only one sort of model. Nonetheless, it is interesting to note the impact of both representational media and the broader cultural environment on patterns of cultural longevity and replicability.

Capacity

Another representational advantage that artefactual media offer information is their far-greater capacity for information than the human memory will ever have. Natural language coevolved with the human brain, and with the medium of speech, for the communication of the content of our thoughts. By definition, therefore, its semantic field is identical with the thoughts that we can hold in our heads. This is a large field, but by definition, too, it excludes any information that we cannot conceptualize without artefactual props. In other words, there is a limit to what we can represent and communicate using natural language alone. There are subjects that we cannot teach without the props of artefactual language; questions that children do not ask until they have the conceptual equipment that these languages provide.

Capacity: Swap Space
One of the ways in which these artefactual languages help is by enabling us to work efficiently with much larger chunks of information than would otherwise be possible. An external representational system enables you to hold a large overall structure of information while simultaneously tinkering with the details. If you only have access to internal representations then your cognitive and conceptual powers are much more limited, because every time you change your focus from one detail to another you run the risk of losing sight of the neglected parts. Artefactual languages do the work of preserving the whole picture while human brains focus on the details, and they can also preserve the details while human brains survey the whole picture.

There is a good analogy for this process in a computer's use of swap space. What happens when you run a program is that the computer reads it from disk into memory (random-access memory, or RAM), where it can actually run the program. The program will almost certainly need to allocate more RAM, either as temporary storage while it performs calculations or as somewhere to store the results of those calculations. If the need for RAM were to exceed the actual size of RAM installed, then the computer would just stop. To get around this, some of the data in the RAM, which are not needed currently or have not been accessed for some time, are dumped out to a special part of the disk called the swap file. This allows the computation to continue. The disadvantage is that reading or writing data to a disk is much slower than reading or writing data to RAM. In fact, there are usually two or three different levels of RAM (an on-central-processing-unit [CPU] cache, an off-CPU cache, and the main RAM), which are increasingly larger and slower. The best performance will be achieved while performing operations that can remain in the smallest and fastest of these levels (the CPU cache). Conversely, if you are trying to run a particularly memory-hungry application on a machine with limited physical RAM, then you'll hear the disk clattering as it works hard to swap data in and out of RAM, and the machine may even grind to a halt.[2]

In normal human cognition, one of the ways in which we use artefactual languages is much the same as the way in which computers use swap space. We dump information that is not currently needed, or has not been accessed for some time, into external artefactual representations. This allows us to think in detail about the parts of the problem with which our brains can cope at one time, or to think holistically about the larger picture that the parts comprise. We can then swap information in and out of artefactual representations, focusing on different parts at each time while preserving the whole in an external cache. Like the computer, we cope best when we are thinking about things that we can hold in our heads, and sometimes grind to a halt when we have inadequate external support for complex computations.

Capacity: Scaffolding

Artefactual languages offer an "economy of storage" (Professor If Price, personal communication), which provides not only the capacity for human minds to deal in more information than their innate memory space would allow, but also a sort of scaffolding for our thinking. Scaffolding can be defined as "a broad class of physical, cognitive and social

---

[2] Thanks to Keith Distin, personal communication, for this explanation.

augmentations – augmentations which allow us to achieve some goal which would otherwise be beyond us" (Clark 1998a: 163).

The concept of scaffolding has an honourable history. Charles Darwin himself asserted that a "complex train of thought can no more be carried on without the aid of words, whether spoken or silent, than a long calculation without the use of figures or algebra. It appears, also, that even ordinary trains of thought almost require some form of language." (Darwin 1871: 57) In the 1930s, the Soviet psychologist Lev Vygotsky (1986) was a pioneer of the theory that cognitive development is profoundly affected by public language, and he suggested in particular that natural language provides scaffolding for our behaviour, as, for example, when a child uses internalized maternal instructions to help him to tie his shoelaces by himself. More recent research confirms that "self-directed speech (be it vocal or silent inner rehearsal) is a crucial cognitive tool that allows us to highlight the most puzzling features of new situations, and to direct and control our own problem-solving actions" (Clark 1998a: 164; see also Berk and Garvin 1984).

One of the most restrictive features of natural language, however, is that it is represented in the serial medium of speech. Some artefactual languages have the capacity to represent information more holistically. A map, for example, represents the geography of a particular area in a way that simultaneously presents all of its informational content to the reader. A drawing, as Turgenev's character Bazarov famously said (1861: 159), "shows me at one glance something that takes ten pages of text to describe." Even a simple graph can present information more efficiently, in a way that is more suited to its content, than the most sophisticated words.

Once again, we can see the intimate coevolution among the content of artefactual languages, their structure and their media: artefactual languages evolve suitable features for representing their particular semantic field in a way that natural language cannot easily capture. Semantic content, linguistic structure and the physical properties of cultural media all combine to provide the framework for what Clark (1998a: 179) calls an "extended mind." He highlights the human tendency to rely on environmental supports such as pen and paper, slide rule, and so on (Clark and Chalmers 1998) – supports that are, of course, representational artefacts. In this way, artefactual languages provide not only swap space – somewhere to dump information while we work on something else – but also the very tools of thought. "Extended intellectual arguments and theses are almost always the product of brains acting in concert with multiple external resources. These resources enable us to pursue manipulations and juxtapositions of ideas and data which would quickly baffle the un-augmented brain" (Clark 1998a: 173). The use of artefactual languages, or in Clark's terminology, "the real environment of printed words and symbols" is crucial in enabling us "to search, store, sequence and reorganize data in ways alien to the on-board repertoire of the biological brain."

Accuracy: Interpretation versus Expression

As well as offering a more persistent and capacious medium than speech, artefactual methods of representation can preserve cultural information with greater accuracy and interpretative precision than the spoken word. One reason for this increased accuracy is that highly repetitive information is subject to less replicative error than information that we encounter only once. So one reason why written instructions are more accurately replicated than verbal information is that they can be consulted as many times as the copier needs, in order to ensure the accuracy of the replicative process (Eerkens and Lipo 2007: 248).

Another reason for their greater accuracy is structural. Because synonyms are avoided in languages that are primarily expressive, whereas homonyms are avoided in languages that evolve primarily for the purpose of hearers being able correctly to interpret the incoming message, Hurford (2003b) and others have argued that the scarcity of synonymy in natural languages, in contrast to the frequency of homonymy, implies that their primary purpose is expressive rather than interpretative. Conversely, a lack of ambiguity is essential in a language whose primary purpose is to be correctly interpreted by its readers. Multiple representations of the same information are fine, but multiple possible meanings of the same representation are not. We can see this principle at work in DNA: the genetic code has redundancy but no ambiguity. Two codons can code for the same amino acid, but the same codon (if it is read in the same direction) cannot code for different amino acids.

Computer languages avoid homonyms for the same reason (Hurford 2003b: 450). In fact, I predict that (subject to this chapter's provisos about cultural context, described in relation to music) analysis of all artefactual languages would reveal that synonymy is relatively frequent, in contrast to the relative scarcity of homonymy. In other words, a written system of representation is more likely to include several different symbols that carry the same meaning than it is to include a polysemous symbol (which carries several different meanings). The preference for synonyms over homonyms in artefactual languages would imply that such systems evolved for primarily interpretative purposes. It would tell us that we are primarily readers – interpreters of artefactual languages – and only secondarily writers, or producers of these written systems of representation. This is in contrast to our use of natural language, of which we are primarily speakers (producers of the language) and to which we are only secondarily listeners (interpreters of it).

Music

Within musical notation, for example, there are synonyms for a note's pitch (A#, B♭), length (♩, ♪ ♪) and expression (·, staccato), but there are no truly polysemous symbols.[3] Any given position on the staff can be used to represent differently pitched notes, it is true, but the clef is always written at the beginning of the staff to clarify which system is being used. Although there are symbols that, if isolated from their context, look identical to each other – a dot can be used either to indicate staccato or to adjust the length of a note by half, for example, and a convex curve can be a phrase or a tie – the context and placement of such symbols serve to disambiguate their meaning. Nevertheless, an interesting facet of musical notation is the extent to which interpretative flexibility does vary across different notational systems. There is a difference, which Charles Seeger (1958) has delineated, between prescriptive and descriptive music writing.

The system of musical notation in Western scores is prescriptive: it provides the key musical ingredients and some suggestion of what to do with them, but there is much in the music that is not notated because it is culturally assumed, and the performer is left to add it in for himself. Some composers, like Johann Sebastian Bach, are famously sparse in their provision of interpretative symbols and consequently draw performers to very different interpretations of their pieces. More generally, performers cannot take music that is represented in Western staff notation and make it sound as the composer intended unless, in addition to an understanding of that system of notation, they also have an understanding of the aural tradition that is associated with it. "For to this aural tradition is customarily left most of the knowledge of 'what happens between the notes'" (Seeger 1958: 186).

---

[3] Many thanks to Rebecca and Andrew Berkley for their contributions to this section, which have greatly increased my understanding of the processes of musical notation and interpretation.

What is left out by the various styles of musical notation, rather than what is included, will reflect the composer's or transcriber's assumptions about what players will bring to their interpretation of the piece. Musicians might have principled views about how music ought to be played, as well as knowledge of the composer's usual style, and this will lead them, for instance, to play a much shorter staccato note in a piece by composers like Bartok and Stockhausen, whose inspiration was drawn from percussive ideas, than they would play a similarly marked note in a piece by composers from a more lyrical tradition, such as Chopin and Debussy.

If we were to try to use this system of musical notation to describe the music of other cultures, then, as Seeger (1958: 186) has put it so aptly, we should need to "do two things, both thoroughly unscientific." First, we should need to select what appeared to us to be familiar structures in the music and write these down, to the exclusion of those structures for which we do not have symbols; and secondly, we should need to expect the resulting symbols to be read and implemented by performers, regardless of their own musical traditions. The results would doubtless be unsatisfactory and this is why, when Western ethnomusicologists transcribe the music of other cultures, they tend not to use traditional staff notation but rather to employ alternative, descriptive systems of musical notation, which aim to offer complete interpretative precision and capture the music in the fullest possible detail. Their descriptive efforts are aimed at ensuring that people from any culture will be able to read and play this music, regardless of their own ignorance of its cultural background.

Like any other language, a system of musical notation is a cooperative game that can be played only by people who are familiar with the shared rules. Meticulously descriptive notational systems ensure interpretative consistency for anyone who is fluent in reading the notation. Prescriptive systems, on the other hand, depend for their interpretation not only on notational fluency but also on cultural familiarity. In this sense they are more like natural languages, whose interpretation also depends on cultural familiarity. For instance, in order to grasp the extent of a visiting American's embarrassment when she realised why her English hosts were so disconcerted by her talk of "khaki pants," we need not only to know the different meanings of *pants* in U.K. and U.S. English, but also to tune into the implications of the different pronunciations of *khaki*, as well as to be aware of how the average Briton might respond to this misunderstood conversation.

Do the more prescriptive notational styles invalidate my claim that artefactual languages, on the whole, avoid homonyms? If musical notation is
sometimes interpretatively ambiguous, so that a quaver in Bartok's score
indicates a note of a different length from a quaver in Debussy's score,
and a note of a different length again when it appears in the final bars of
the Debussy, when a player will often be slowing down as the piece draws
to a close, then doesn't this mean that the quaver is a polysemous symbol?
Doesn't it mean that all musical notation is effectively polysemous, with
its precise meaning determined by the context in which it appears? How
is this any more accurate than natural language, whose meaning is also
largely context dependent?

These questions can only be answered once we understand that the
representational accuracy of any language will be determined, not by
whether it is natural or artefactual per se, but by the extent to which its
receivers can be assumed to have the cultural knowledge that is needed to
disambiguate any homonyms. In general, artefactual languages will need
to have more interpretative precision than natural languages, which have
evolved for communication between social groups, but musical notation
is different, in this respect, from most other artefactual languages. Musical notation, like natural language, has evolved for the representation of
information that is culture specific.

The purpose of natural language is to enable a particular group of people to communicate with one another, and its interpretation depends on
(often unspoken) culture-specific knowledge in addition to fluency in
the particular language. For example, any competent English speaker,
listening to a story being read aloud, would easily be able to understand
this extract: "The chief engineer looked angry. *Don't do that*, she said."
A member of the audience who is not familiar with English culture as
well as with the English language might not, however, do the slight mental double-take that many English listeners would do when they heard
she in this context. Their reaction has nothing to do with the fluency
with which they speak English and everything to do with the cultural
assumptions that they bring to the story. Similarly, the purpose of musical notation is to enable a particular group of people to make music
with one another, and the interpretation of prescriptive systems of musical notation depends on (often unspoken) culture-specific knowledge
in addition to fluency in the notational system. These notational systems
are founded on a set of culture-specific assumptions about what people
will bring to their musical interpretation: the ways in which they will produce a performance based on that written music. Such assumptions can
be made because, like natural language, musical notation has evolved

for use within a specific culture. This is why, for Western ethnomusicologists whose own cultural background invalidates those assumptions, a minutely descriptive notation becomes necessary.

Moreover, we should be careful not to confuse even the most complete visual representation of the music with the full sensory and perceptual reaction of a listener who has been conditioned by the music-cultural tradition of its origins. "The physical stimulus constituted by a product of any music tradition is identical to those who carry the tradition and to those who carry another. It is the perceptions of it by the respective carriers that are different" (Seeger 1958: 195). Music, like any other cultural product, is dependent on the nature of the receiver for the way in which it is interpreted and experienced.

Languages of all kinds coevolve with their semantic fields: the purpose of musical notation is to enable musicians from within a particular culture to play that culture's music, and each notational system is almost bound to have a more limited potential for representing the music of other cultures, with which it has not coevolved. Thus, the incompleteness of prescriptive musical notation is a result of its location within a particular cultural context. It evolved for use within a particular group of people, and therefore it can rely for disambiguation on the cultural knowledge that those people bring to it. Where this cultural knowledge cannot be assumed, a more precisely descriptive system is needed to avoid ambiguity. Representational pressure towards interpretative unambiguity is greatest when those who share a representational system cannot be assumed also to share the relevant portion of disambiguating cultural knowledge.

Dialects of Artefactual Languages

Another source of interpretative ambiguity, to which some artefactual languages are subject, is the emergence of local variants, which represent the same information in slightly different ways. Spellings in written English, for example, vary with the author's country of origin: common examples include *colour* and *color*, *centre* and *center* and *cheque* and *check*. Other artefactual languages that show dialectal variations include mathematical notation and methodology, and the conventions of orthographic projection, which is a method of showing what an object looks like from different directions.[4] The American style of presenting the different views, which is known as third-angle projection, shows the viewer

---

[4] In recent years, the use of three-dimensional object manipulation by computer-aided design packages has largely dispensed with the need for such drawings.

what he would see if a sheet of glass were placed over the object and the object rotated under the glass. Thus, with an object's front view in the centre of the page, the view of its left-hand end would be drawn to the left of that front view, and its right-hand view would be drawn on the right. In the European style of first-angle projection, however, the viewer is shown what he would see if the object were rotated on top of a table, with the (perhaps counterintuitive) result that the view of each end of the object appears on the opposite side of the drawing from its "own" end of the front view. Both projections result in the same views, but they arrange them differently on the page. For this reason, a symbol is added to the drawing to indicate which angle of projection has been used: a truncated cone drawn in the appropriate angle of projection.

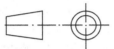 

Orthographic First-Angle Projection   *Orthographic Third-Angle Projection*

Like natural language use, the use of artefactual languages for representing information is a cooperative game played according to shared rules. As a result, any variation in those rules can produce a corresponding difficulty for the players: a lessening of the ability to make shared sense of what is written (or otherwise represented). Given that artefactual languages are representational systems, any unexpected difference in the vocabulary or structure of the representational symbols is bound to interfere with the production of meaning, at least at first glance. Moreover, in the absence of any indication of which local variant is being used, artefactual language dialects can give rise to ambiguity. Without the appropriate symbol, for instance, the layout of an orthographic first-angle projection could be deeply confusing for an engineer who is used to third-angle projections. Without any indication of the author's nationality, what are we to make of his claim to wear suspenders to work every day? Local variants have the potential to introduce an unacceptable degree of homonymy into representational systems whose primary purpose is interpretative.

For this reason, noticeable differences in artefactual language use, including some that are trivial enough to preserve meaning, act as important badges of *representational* identity. When we are reading written English and see the words *color* or *cheque*, they tell us which local variant is being used, and this knowledge disambiguates later statements about businessmen who go to work in pants and suspenders. The way in which we use artefactual languages – whose primary purpose is representation – tells us which representational system we are using.

In contrast, the way in which we use natural languages – whose primary purpose is communication – tells us which group of communicators we belong to. Noticeable differences in natural language use, including some that are trivial enough to preserve meaning, act to identify us as members of the same group. If you speak my language with my local accent, using dialect words from my local region, then this tells me all sorts of other things that we are likely to have in common. As well as enabling us to communicate with each other, it also marks us out to each other as members of a particular social group, and changes our attitude towards each other. In other words, local natural language variants act to disambiguate social relationships. The way in which we use natural language – a medium whose primary purpose is communication – is a badge of the group in which we learned to use it.

Local language variants, then, can play both interpretative and communicative roles, disambiguating (or, conversely, excluding us from accessing) meaning as well as affecting our attitude towards the language user. The different weight that we give to these effects in each case is a result of the different functions and origins of each kind of language. Because the primary purpose of natural language is communication, the identification of a speaker's natural language or dialect is often most useful as a marker of whether communication will be fruitful or even possible. It is of little interpretative use for an individual to identify a language or dialect other than her own, if it is not a variant with which she is familiar: an English monolingual might be reasonably able to identify when somebody is speaking German, but this will not help her to understand what he is saying. Where such identification is of use is in delineating the social ties that might exist between the language users.

Artefactual languages do not normally have these strong local links to a particular group, largely because they are not the means by which infants learn to communicate with other group members. Where artefactual languages are culture specific, like prescriptive systems of musical notation, this is because they have coevolved with the culture-specific nature of their content. Staff notation has evolved as a receptacle for the music of a particular culture, and has limited utility outside that culture's shared assumptions. Although all artefactual languages coevolve with their content, most of them do not have content that is culture specific in the way that music is, and for this reason their representational systems tend also to be culture neutral. Because these artefactual languages have to produce a complete description of their content, rather than relying on their interpreters' shared cultural heritage to carry a portion of it, they can be used by anyone who is fluent in that artefactual language, regardless of her cultural background.

Artefactual language variants may sometimes bring communicative as well as interpretative information: the first-angle projection symbol does not only tell me how to interpret an engineering drawing; it also identifies the work context as European rather than North American. Nonetheless, this does not bring me reliable information about the cultural origin of the drawer, because he will be forced to use the system that is appropriate to his work context, no matter which he happens to have learned first. Artefactual languages are acquired throughout our lives: we are not dependent on the innate acquisition of one artefactual language during childhood, but can pick up new ways of representing information as they are encountered or needed. For this reason, information about the identity of an artefactual language or local variant does not provide reliable information about the user's social background. Unless there is a firm link between an artefactual language or dialect and a particular social group, and a person is already familiar with this link, then the identification of that artefactual language or dialect is of little use in delineating the social ties that might exist between himself and other users of that artefactual language. An English engineer might be able to identify which angle of orthographic projection a drawing is using, but this need tell him nothing about the background of the drawing's originators.

It would, however, tell him how to interpret the drawing's meaning, and this would enable him to work with its originators. The shared use of an artefactual language enables people to get things done together, even in the absence of any social ties, as the next section describes in more detail.

Detachment

The stability, capacity and accuracy that artefactual media provide for cultural information strongly suggest that artefactual languages evolved for representational, rather than communicative reasons, and add weight to the theory that they can act as mechanisms of persistent cultural heredity, and hence of cultural evolution. Whereas the spoken language evolved as an efficient means of reproducing information, artefactual languages are particularly well suited to preserving cultural information in a lasting and unambiguous format. The other adaptive advantage that they offer is their separation from the humans who use them. In the absence of artefactual media, we cannot communicate unless we are within sight or sound of one another, for our words, like the gestures of sign language, are a physical part of us. As a result, such media are useless as a means of transmitting information between people who are distant from each other, whether physically, socially or temporally.

Artefactual media, on the other hand, can be used to transmit information between people who might never even meet. An artefactual language can be used for communication over time, space and social barriers, bringing benefits both to its users and to the information that it represents. This is true even of the earliest systems of artefactual representations such as clay tokens, for example, which could be shared by people who spoke different natural languages (Schmandt-Besserat 1996: 83).

Natural language has evolved as an innately acquired means of communication among members of a community. It forges and strengthens the social ties between members, increasing their ability to find mates, to establish a place in the social pecking order, and to cooperate and agree on shared systems of values and mores. So vital is it to these processes that it has coevolved with the human brain to the extent that children can learn it instinctively. But there is also value in people being able to work together, even when they do not have any social links with one another. This can be particularly useful when their task is complex or labour intensive. An artefactual language, in these cases, can form a bridge between people who do not share any other connections, because it provides a means of exchanging information about the task, even in the absence of any means of communication about anything else. Indeed, as we saw in the evolution of musical notation, the representational pressure towards interpretative unambiguity is greatest when assumptions cannot be made about the cultural backgrounds of those who will be using a representational system.

The distinction between social and functional ties is as relevant today as it has ever been throughout history. In contemporary work settings, for example, we often need to decontextualize and depersonalize our interactions: when dealing with government bureaucracy, banks and so on, we find that our background and personal experience, our natural emotional responses or social intuitions are worthless, and instead we need to apply the rules and do business with each other, rather than engaging socially. "The worker must realize that he/she is not in an *actual* social interaction with this person . . . but in a special, indeed 'unnatural' realm where different rules apply" (Stanovich and West 2003: 15).

Education, Technology and Development

If it is true that artefactual languages facilitate functional links between their users, then we should expect to find empirical evidence of a causal relationship between a society's adoption of an artefactual language and the ways in which its members cooperate with other people, both within and beyond that society. The widespread acquisition of artefactual languages like writing and mathematical notation should be predictive of their receivers' capacity to work with others outside their kinship and social groups, and this in turn might be expected to bring economic advantage. We should perhaps expect to find a correlation between levels of literacy and education, and levels of both social and geographical mobility. Members of preliterate societies, on this theory, would not be expected to form connections outside their kinship and social groups, whereas more technologically advanced societies should have members whose livelihoods are not dependent, to the same extent, on the quality of their relationships, because they can work with other members even when they have nothing else in common with one another. "Their natural milieu," as Durkheim (1964: 182) put it, "is no longer the natal milieu, but the occupational milieu. It is no longer real or fictitious consanguinity which marks the place of each one, but the function which he fulfils."

Certainly, there is empirical evidence of a statistically significant, positive relationship between income and increases in the literacy rate (Andreosso-O'Callaghan 2002: 4). Modern economic growth is accelerated by the establishment of modern education systems (Easterlin 1981), in ways that reflect the relationship between artefactual languages and their content. Richard Easterlin (1981: 3) has shown how international differential economic growth during the past two centuries is a product of the accelerating rates of technological change in relatively few nations. Easterlin asks why there should have been such a constraint on the dissemination of this new, modern technological knowledge. His answer is that "the more schooling of appropriate content that a nation's population had, the easier it was to master the new technological knowledge becoming available" (1981:6).

A substantial schooling system is, it seems, a prerequisite for sustained economic growth, and the quality of education on offer must have some influence on this relationship – but this is not the only factor at play. Bernadette Andreosso-O'Callaghan's (2002) analysis of ten Asian economies, between 1980 and 1997, revealed that as well as educational *quality*, different *levels* of education will have different effects on economic growth, depending on each country's level of economic development. In developed countries, there is of course a two-way causality at work, with economic growth explaining why countries can afford to develop tertiary education, and tertiary education partially explaining that economic growth. In less developed countries, it is primary education that contributes most to economic growth, by enabling pupils to understand and use simple technologies, and by leading to macroeconomic gains in productivity.

In either case, we can see that what is needed is the right level of education for the level of cultural information that is available: the coevolution, in other words, of languages and their content. Economic growth is founded on the ability of people to work together, and this ability is dependent on individual participants' grasp of the relevant cultural languages. Some tasks can be achieved by people who speak the same natural language; others depend on the ability to use artefactual languages like writing or mathematical notation, if only so that the people involved are able to grasp the concepts that these artefactual languages facilitate. The population of any country needs to be educated to a level that allows it to participate in the wealth-creating activities that go on in their country. More complex technologies and social institutions become affordable as wealth increases, and if there are to be workers to engage in these activities then the levels of requisite education will need to rise accordingly. Education transmits both artefactual languages and the information content with which each has coevolved, to a population that is thereby enabled to access this content and make functional links with one another, working together on wealth-producing projects that they would have been unable to achieve without the benefit of education.

Functional Links

In the competition for human attention, there are clear adaptive advantages for information that finds a way of being preserved over time, conveyed over space or transmitted over social barriers without relying on the life span, peregrination or sociability of the person who produced it. In addition to the advantages that it brings to the information that artefactual languages represent, this detachment brings obvious benefits to the humans who use them. In the same way that speech provides and strengthens the social links between members of a community, artefactual media can provide functional links between people whose communities are separated by time, space or social divisions. Artefactual languages are not acquired innately in infancy, do not shape group members' perceptions of other speakers and are not the basis of communication within the group. It is for this reason that they can form cultural bridges between otherwise-unrelated groups of people.

If you cannot speak my language but are nonetheless able to sit in the orchestra with me, reading music from the same score and interpreting the conductor's gestures in the same way that I do, then we might not be able to communicate very well, but we can nonetheless make music together. Given the necessary skills and physical capacities, if you know the rules of hockey then you can join my team and we can play a game together; if you understand the conventions of my engineering drawings then you can build the road that I have designed; if you are able to use the same currency as me then we can trade with each other. Artefactual languages thus provide functional links between people who are otherwise unrelated, enabling them to get things done together even when they are members of different social groups.

Natural languages have evolved for communication and act as an effective marker of social identity, indicating the social links that exist among their users. Artefactual languages have evolved for representation and act as an effective marker of cultural identity, indicating the functional links that exist among their users. Competence in an artefactual language provides cultural access to a particular group, just as competence in a native natural language provides social access to a particular group.

This is because, despite the much-greater degree of interpretative precision that artefactual languages offer over natural language, there is an extent to which artefactual languages make assumptions about the shared knowledge that readers will bring to them. The conventions of engineering drawings, for instance, will specify all functionally significant features of a design but may leave presentational or aesthetic decisions open to the user. They assume that the reader understands why some aspects of the design have been precisely specified and other elements have not. The use of an artefactual language is a cooperative game just as the use of a natural language is, and if you do not know the shared set of rules then you are excluded from the group of people who use that artefactual language.

This exclusion is not selectively critical in the way that the failure to learn the communication system of your local group would be, but it will have implications for your role and status within society, as well as for your potential to move between groups. In this way, there might well be a degree of correlation between biological fitness and the use of some artefactual languages. Your exclusion from the group of people who can use that artefactual language will not prevent you from communicating with them on a social level, but it will exclude you from knowledge of the semantic field represented by that artefactual language, from the activities in which that artefactual language enables people to engage, and from knowledge of the standards against which members of that cultural group are judged. All of this may well, of course, have social implications, as the following chapters explore.

9

Money

An Artefactual Language

Culture is the product of interactions between human receivers and cultural information, which is represented in a variety of culturally evolved natural and artefactual languages. Natural language may have evolved primarily for communication, but the eventual result was an expansion in cultural information beyond the collective capacity of human brains. At that point, our ancestors began to use artefactual symbols as permanent, external receptacles for cultural information. These artefactual symbols depended primarily on the clever trick of human hands, just as the natural language symbols had depended primarily on the clever trick of human voices (a view that was developed in personal communication with If Price, 2006). Symbols are only of use to a community if everyone is using the same system; but, for reasons that I have sketched in Chapter 7, artefactual symbols could not rely on coevolution with biologically endowed cognition to facilitate their acquisition by our ancestors. If they were to succeed as communal systems of representation, then they would need to be explicitly teachable rather than innately learnable. As the number of symbols expanded, therefore, the bottleneck in transmission between writers and readers ensured the emergence of compositional rules governing the artefactual symbols.

Artefactual languages enable cultural information to be preserved for greater lengths of time, in greater quantities and with greater accuracy than natural language can achieve. They provide scaffolding and swap space for human cognition, and create functional links between different social groups. In these ways, they have propelled cultural evolution by providing a mechanism for the preservation and inheritance of increasing amounts of cultural variation. Variations in cultural information are inherited by human receivers who have learned the language in which it is transmitted, and the same abilities that enable humans to acquire artefactual languages also enable us to receive the information that they represent. Artefactual languages thus provide a vital mechanism for modern cultural evolution.

The evolution of money provides a particularly clear example of the way in which artefactual languages coevolve with their content and media, and provide functional links between socially disparate groups, with potentially far-reaching social implications. A common criticism of cultural evolution theories is that they have little explanatory value, but in this chapter and the next I present evidence that money is the product of cultural evolution, and demonstrate how this account contributes to our understanding of the monetary phenomena that are studied by disciplines like economics, sociology and psychology. In this chapter, I explore the cultural evolution of money as an artefactual language for the measurement of value, which is realised in one or more media of exchange, and creates functional relationships between the people who agree to measure and exchange value in this format. I argue that money derives the value that it measures from this collective agreement, and I emphasise the interdependence of its functions as a measure of value and a medium of exchange. In the following chapter, I explore the explanatory value of this representational account of money.

Money and Barter

Humans are social creatures who depend on one another in many respects. When individuals cannot provide certain goods or services for themselves but are members of a community, operating within the constraints of shared customs and morals, and possibly even related to one another by blood or marriage, barter is one of the most successful systems of exchange. It might involve the exchange of goods like grain, livestock, fish or tools, or the exchange of services like babysitting, holding the key to a neighbour's house or assisting with practical tasks. Barter works best in bilateral exchanges between people who want what each other has to offer, and who trust each other to work out a fair way of comparing the values of the goods or services that are being exchanged.

Barter is not an ideal system for every circumstance, however. It is cumbersome, because one person does not always want what another has to offer in return for his products or skills. It is also limited in use, because it provides no common standard of value for a variety of goods and services, and is therefore dependent on the existence of a web of social ties among participants. When commodity owners are essentially "foreign" to one another, with nothing in common except the economic exchange in which they wish to engage, there is no guarantee that one commodity owner will accept another's commodity in exchange for his own, or that a mutually agreeable rate of exchange can be fixed for the barter (Lapavitsas 2003).

Some sort of social nexus is needed among "foreigners" who meet often and in large numbers for trade purposes. This nexus is money: a discretized measure of value, "which makes it possible to establish the relative prices of all commodities" (Ingham 2008: 68). Moreover, it is not only the problems of commerce among foreigners that money can solve. There are other circumstances in which members of the same community can become, in effect, foreigners to one another: in which the ties between them are broken and a new means of connection is needed. In early societies with laws that demanded compensation for crimes of violence, for instance, peace was made via payments that had, of course, to be in a unit of value that both sides could accept. The same is true of bride money, which compensates the head of the family for the loss of a daughter's services; it is true, too, of taxes and of religious tributes. Cultural exchanges that had nothing to do with commercial trade were nonetheless dependent on a common unit of account for the calculation of debts and credits (Davies 1997).

Even such a brief sketch of the difference between money and barter focuses our attention on the three distinctive and interdependent properties of money that I shall be highlighting in this chapter. Barter is an essentially bilateral process, which relies on existing social relationships among participants to establish an agreed exchange value, for goods and services that may or may not be wanted on any given occasion. In contrast, money facilitates multilateral transactions, because a *functional relationship* is created among people when they all agree to *measure value* in the same way, and will all accept the same *medium of exchange.*

The Medium of Exchange

Lapavitsas asks us to imagine two commodity owners who approach each other as foreign individuals, concerned only with obtaining something of equivalent value to what they have brought to market. When one of them makes an offer to sell what she has, she effectively requests exchange with the other and declares that the exchange value of her own commodity is a certain quantity of the other owner's commodity. The buying power of this other owner's commodity derives entirely from the first owner's offer to sell.

Each commodity owner can offer to sell her product to many potential buyers, and this has the potential to confer buying power on many different commodities. What is unique about money, according to Lapavitsas (2003: 13), is that it "represents the concentration of the ability to buy in one commodity." Lapavitsas supports the modern anthropological view that money is not merely a more convenient substitute for bartering, but rather arose in the context of prevailing social customs and priorities. For example, commodities like cattle and grain were of inherent value in early society, which meant that they were desirable by almost every other commodity owner and could be exchanged for a broad range of other goods and services. Over time, such commodities developed a dual role as both usable commodities and units of exchange that were perceived as a common standard of value. In this way, Lapavitsas argues, money could have emerged spontaneously from the developing network of social and economic relationships between foreign commodity owners. Monetary exchange was shaped by existing social patterns, but it is qualitatively different from barter because money "is the only commodity that can buy all others" (Lapavitsas 2005a: 26).

Lapavitsas's account highlights both "the social character of money's emergence" and the physical characteristics of the commodities that have come to monopolise direct exchangeability. Precious metals like gold and silver, for example, had the advantages of a traditional association with wealth display, as well as physical properties that made them particularly suited to the role of money: metal is both durable and easily divided, for example – unlike a cow. On the other hand, metals also had the disadvantage of depending for their acceptance on being weighed and assayed. It was for this reason that pieces of metal began to be stamped with a mark that ascertained first the quality and later both the quality and the weight of that piece of metal. Thus, the lengthy and fallible processes of weighing and assaying were replaced by the simpler process of counting. The first coins of this sort were made of electrum, an alloy of gold and silver, and were introduced by the Lydians in the seventh century BC.

In summary, Lapavitsas provides an account in which money has its roots in the need for "foreign" traders to forge useful connections with one another. Initial contacts could have led to the exchange of virtually any one commodity for another, but over time there emerged a natural

preference among sellers to exchange their goods and services for commodities that were highly valued by members of their own communities.
Among these commodities, the ones that were easily divisible, transportable and durable were particularly favoured. In this way, metal emerged
as the monopolist of buying power in some communities, whereas grain
gained the monetary crown in others.

The Measurement of Value

Both of these commodities could be discretized in the way that was
needed if money were to act as a measure of the value that was being
exchanged. This is important because all forms of money must be able
to function both as "monies of account" and as "monetary media" (Dodd
2005: 559). Whatever the chosen monopolist of buying power, it was
useless as a medium of exchange unless there was a way of measuring the
relative value of each portion of it. An innumerate child, for example,
may have a basic understanding of money as a means of exchange, in
the sense of knowing that money is something we need in order to
acquire the goods that we'd like, without having any comprehension of
the amount of money that is needed to purchase any particular item. To
our younger son at the age of four, for instance, there was no difference
between a pocketful of pound coins and a pocketful of pennies. If, as
quite frequently and incomprehensibly happened, what he really wanted
to buy was a small egg made from a type of chocolate that he disliked,
containing a tiny and inescapably disappointing plastic toy, then he would
not have cared whether you gave it to him in exchange for ten pennies,
ten five-pence coins or ten pound coins. Nor, conversely, was he able to
understand that the pile of copper coins in his money box was not, despite
its appealing weightiness, sufficient to purchase the dressing-up outfit
that he so desired. His understanding, at that stage, had not progressed
beyond the knowledge that goods and services are exchanged for "some
money." In other words, although he knew that money is a medium of
exchange, he did not yet understand how it measures value, and this
meant that he was effectively excluded from the use of money.

The importance of money as a measurement of value has been emphasised by Geoffrey Ingham (2008: 68), who criticizes Lapavitsas's account
of the history and nature of money, on the grounds that barter exchange
is neither theoretically adequate for the production of a universally equivalent commodity that can act as a stable measure of value, nor indeed
implicated in that role by the evidence from anthropological history.

Whereas barter is a bilateral exchange, for which agreement between the two sides is sufficient, the multilateral exchange of heterogeneous goods presupposes a stable and universal measure of value (Ingham 1996: 526), without which traders cannot establish relative prices. On Ingham's view, money is not an exchangeable commodity but "primarily a measure of value and a credit relationship" (1996: 515): currencies measure a credit-debit relationship in which money's issuers (e.g., banks, states) promise to accept that money in payment of any debt owed to them.

The debate between Lapavitsas and Ingham is conducted in the shadow of larger theoretical issues than can be accommodated here. Looming behind their conflicting theories of money as a universal equivalent, and money as a measure of abstract value, are the spectres of Marx and Keynes, among others, and I am not qualified to exorcise these ghosts. Thankfully, however, there is no need for me to venture behind this theoretical backcloth, because the theory of money that I am developing in this chapter is spun of strands drawn from several alternative views.

Lapavitsas and Ingham agree that a medium of exchange cannot function as money unless it is also capable of acting as a measure of value, but they disagree about what gives the medium its value. For Lapavitsas (2003: 15), "the functioning of money as a measure of value cannot be separated from its functioning as a means of exchange." Its value derives from its universal desirability *by commodity owners*: a commodity becomes money when it can be used in exchange for anything, because it has an agreed amount of value to commodity owners. For Ingham (2004: 12), however, money "cannot be understood without reference to an authority," for its value derives from the credit-debit relationship between its issuers and its users: a commodity becomes money when it can be used as payment of any debt to the issuing authority, because it has an agreed amount of value *to the issuer*.

In order to unpick these two approaches, it is useful to note that Ingham effectively identifies money with state-issued currency (Dodd 2005:561), whereas Lapavitsas's definition embraces a range of monetary forms, of which state-issued currency is just one "historically specific entity" (Dodd 2005: 563). So the interesting question, here, is whether Lapavitsas has mistakenly included some nonmonetary exchange media in his definition of money, or Ingham has mistakenly excluded some monetary exchange media from his definition of money.

I am persuaded by Lapavitsas that the evolutionary imperative for money was provided by the need for functional links among people who were literally foreign to one another, or whose social circumstances had effectively estranged them in the process of bringing them into contact with one another. Barter relies on participants' trusting each other to make a fair exchange, and an alternative is needed when social relationships are absent or broken. I would also agree with Lapavitsas's assertion that one of the qualitative differences between money and barter is that money is universally desirable in exchange for any given commodity. But I believe that Lapavitsas may have confused the order of causality that is at work here. It seems to me – and the remainder of this chapter explores this thought in more detail – that a given commodity does not become money because it is universally desirable, but rather a particular commodity becomes universally desirable when it begins to function as money, by virtue of its new use as a measure of value. In this sense, I am persuaded by Ingham that the essential function of money is its measurement of value, which I would describe in terms of its operating as an artefactual language in which value is represented. On the other hand, because any language must be realised in a medium of some kind, I believe (and will explain further) that Ingham has underplayed the need for a monetary medium of exchange.

The Derivation of Value

Both Lapavitsas and Ingham agree that money provides a stable and universal measure of value, and both agree that money derives its value from an agreement between its users to measure value in a particular way. The distance between their positions is primarily created by Ingham's insistence that an issuing authority must be one of the parties to the agreement, and by Lapavitsas's insistence that money's defining feature is its universal desirability as a commodity. According to Lapavitsas, there may or may not be an issuing authority, but money must have its basis in a medium that is universally desired by those who have something to offer in exchange for it; whereas Ingham maintains that money may or may not have its basis in an actual medium of exchange, but what matters is the existence of an issuing authority.

Barter, as Ingham points out, is an essentially bilateral process, in which an exchange is agreed on the basis of existing social relationships and values. Both sides can agree that a certain weight of one commodity is equal in value to a given number of pieces of a different commodity, or to a day's labour on a particular task, because both sides trust each other to act fairly, and they have a shared understanding of what *fairly* means in their society. It is when people need to do cultural or commercial business with those outside their usual circle of trust that the methods of barter break down.

Money facilitates multilateral transactions by providing a measure of value that all sides can accept. Money, as Lapavitsas says, creates a functional nexus between the people who agree to use it in their transactions, even when they have no other relationships with one another. The interesting question, for me, concerns the basis on which this agreement is reached: if money is a stable and universally acceptable measure of value, then we need to know from what it derives its lasting value to its users, and what makes them universally prepared to accept that way of measuring the value that they attribute to it.

Humans who agree to use a particular commodity as the medium of exchange must also agreed to discretize value in a way that matches the measured volume, weight or quantities of that commodity. My theory of money rests on the claim (defended in more detail below) that it is from this collective agreement that the commodity derives its value *as a medium of exchange*. We must not be misled by the fact that this commodity has been *selected* as a medium of exchange because it already had another, established value to the members of the societies in which it is used in this way. When one commodity is selected to be used as money – the universal medium of exchange – it attains a new type of value, in addition to its intrinsic utility, such that its owners can now choose to exchange it instead of using it. The utilitarian value of the commodity (e.g., as food or as a precious metal) is the reason why it was originally selected as a medium of exchange, but its value as a medium of exchange – the particular exchange value of each measured portion of that commodity – derives entirely from the collective agreement of those who have chosen to measure and exchange value in that way. Commodities become money only "when they are further 'described' in terms of the measure of abstract value – pound, dollar, euro and so on" (Ingham 2008: 69). Thus, when a human receives a measured amount of a commodity that has been adopted as money, it is the *measurement* which represents the value of that amount of the commodity *as a medium of exchange*.

Credit Money

Once that system of measurement has been widely adopted, it can begin to evolve in different directions from the medium in which it originated. In this way, when the representational systems of commodity money were applied to new media, various types of credit money emerged. The public stamps that had symbolised the value of the earliest coins provided a guarantee of the quality and weight of the metal contained in each coin. Over time, however, the actual value of the metal that it contained became less important than a coin's symbolic value as a medium and standard measure of exchange. Marks on a cheap piece of metal could serve the same purpose as marks on an intrinsically valuable piece of metal. Unlike the earlier stamps on precious coins, the stamps on the cheaper pieces of metal did not represent the metal they contained: they provided a guarantee that the bearer of the coin could use it to purchase goods or services, or to settle debts, or to exchange it for precious metal, to the value that the marks represented.

Money of this sort, whose value is not matched by its worth as a commodity, is known as credit money. One of the most sophisticated forms of credit money was developed, in conjunction with a national banking system, by the Egyptians in the fourth century BC. Their credit money took the form of receipts issued to farmers in return for any surplus grains that they deposited in royal or private warehouses. These warehouse receipts were an easily transportable, trusted symbol of the grains that they represented. It was even possible for the warehouses to provide loans, by issuing new receipts to people who had not yet deposited any grain, so long as the overall value of the grain in the warehouse was sufficient to maintain the public's trust.

These warehousing "banks" went even further in their use of monetary representations. When Egypt fell to the Greek Ptolemaic dynasty in the 320s BC, the various government granaries were transformed into a network with its headquarters in Alexandria. The accounts of every warehouse were recorded in this central Alexandrian bank, and the numbers that were recorded in these account books became symbols of the money that each depositor held. In this way, transfers could be made, when balances were changed in the account books, without a single receipt being exchanged.

Despite the sophistication of the Ptolemaic Egyptian banking system, the Romans preferred to use coins for the sorts of functions that banks had provided. As a result, banking was forgotten and not reinvented until much later, when it was closely linked to the emergence of paper money. Although the Chinese had been using paper money since the tenth century AD, in the West it was not until the Crusades stimulated the revival of banking that written instructions, known as bills of exchange, began to be used for the transfer of money. Later, during the English

Civil War in the seventeenth century, written instructions to goldsmiths were the predecessors of cheques, and banknotes developed from the use of goldsmiths' receipts as evidence of ability to pay.

Each of these forms of credit money was of use only within the circle of people who trusted the issuer. When a country's government is the issuer, everyone in the country can use that form of credit money in settlement of their debts (this is the defining feature of legal tender), so long as trust is maintained in the government and its economy.

Fiat Money

Although credit money itself has little or no intrinsic value, it is backed by a commodity like silver or gold. On a British ten-pound note, for example, the words, "I promise to pay the bearer on demand the sum of ten pounds," together with the signature of the Bank of England's chief cashier, is a reminder that banknotes were originally regarded, not only in Britain but also in many other countries, as a money substitute, and were not accepted as the real thing until much later (Davies 2005). Once the population's trust in the government and economy of their country is strong enough, however, there is no need for their money to be a credit note. Today, no major modern currency is either backed by or convertible into a commodity. Instead, it is fiat money, backed by the public's trust in the stability of their country's government and economy. Its value is set, not by how much of a commodity it is worth, but by what the government says it is worth.

The qualitative difference between credit money and fiat money, then, is that credit money is a promise to pay, whereas fiat money – like commodity money – is the final payment itself. A given amount of fiat money does not *represent* a payment, in the form of a particular portion of a commodity or the promise to hand over that portion of the commodity: it *is* the payment. The notes or coins of fiat money actually have the value that is represented by their respective denominations.

Does this reduce fiat money to a self-referential system, whose "creative and autonomous" symbols are the only source of reality for the value that they describe (Rotman 1987: 28): a scandalously self-created fiction, printed or minted into existence (Rotman 1987: 50)? At first glance, it does seem that there are significant differences between real-world commodities, like bullion or cows, and the so-called imaginary money (Rotman 1987) of modern currencies, whose notes and coins are not backed by a promise to convert them into a fixed quantity of anything.

Closer inspection, however, reveals that the similarities between different types of money are much more significant than these differences.

The Functional Relationship

Just as every individual in every society values some things more than others, so societies collectively put a different value on different goods and services, the social and physical consequences of individual actions and life events, and so on. In order to deal fairly with one another, societies' members must find collectively acceptable media in which this value can be exchanged; and in order to measure the value of one action or entity against another, they need to find a collectively acceptable way to discretize value. An acceptable medium of exchange will need to be valued by all participants and discretized into units whose size matches the scale on which they wish to make exchanges. The question that was raised in a previous section was whether the medium's discretized units derive their value from their privileged acceptance by an issuing authority or from their universal acceptance by commodity owners. We are now in a position to see that the fundamental similarity between the two options: in all cases, money derives its value from an agreement between its users to represent value in a particular artefactual language and to realize it in a given medium.

In this sense, fiat money derives its value from exactly the same source as credit or commodity money does: a collective agreement. Although it is true that each symbol in a currency (as in any other representational system) derives its meaning from the overall system, and that the system affects how its users discretize value, this should not be taken to entail that currencies create the value that they measure. (After all, the same currency can be used to measure the value of both credit and fiat money: the dollar, for example, was the U.S. currency both before and after the 1971 suspension of its convertibility into gold.) Rather, the humans who adopt a shared currency have agreed to discretize value in a way that matches the denominations of that currency, and to use that currency's coinage as a medium of exchange, and it is from this collective agreement that the currency derives its value.

Just as the measurement of a commodity represents its value as a medium of exchange, so the symbols on the coinage of modern currencies represent the value of that coin or note as a medium of exchange. The fact that the commodity also has a utilitarian value, which the fiat money does not, is irrelevant to the derivation of the value of each *as a medium of exchange*, which in both cases is the product of a collective human agreement. In this sense, commodity money is no less "imaginary" than fiat money. Both derive their value from the agreement of a particular group of humans to discretize value in a particular way and to use a particular medium for its exchange: "all money is 'virtual' and is defined by use, information and networks of social relationships" (Singh 2000: 4.3). Commodity money is of use only to people who share a desire for the relevant commodity – a desire that has been uniquely concentrated on that commodity because they have all agreed on its value – and who trust its discretization as a method of measuring value. Credit money is of use only to the group of people who trust its issuer. Fiat money is of use only for so long as the public trusts its government. In these ways, just as a local natural language provides communicative links between its users and identifies them as members of a particular social group, so a local currency provides monetary links between its users, and identifies them as members of a particular trading group (though monetary links can also be forged for reasons other than trade). Nonspeakers of the local language cannot communicate with group members, and people without the local currency cannot trade with those whose business relies on that currency. Money, in all of its forms, is like any other artefactual language, enabling its users to cooperate with one another in a particular cultural field, and excluding from the circle of cooperation anyone who will not use that language to discretize its semantic content.

Like any other artefactual language, too, the particular form that money takes in any given context – both the representational system that it uses and the medium in which it is realised – is bound to have an impact on its function and evolution. There is a three-level variation in money's properties: the way in which "money integrates into interpersonal relations" both affects and is affected by money's media and its measurements of value (Zelizer 2005: 587). Money, as the next section will show, relies on a representational system that has coevolved with its medium and content, and as an artefactual language it provides links between groups of people who are otherwise unrelated.

Money's Cultural Evolution

Money has three separate but interdependent functions: it acts as a medium of exchange and a measure of value, and as a result it provides functional links between its users. This threefold analysis is crucial to the understanding of its evolution. Money is a beautifully illustrative example of the interplay between the evolution of an aspect of culture, its representational system, the medium in which it is realized and the cultural context in which it emerges.

From a cultural evolutionary perspective, the emergence of money can be explained as a result of the emerging human desire for a way of measuring the value exchanged in various social transactions. As in any other cultural area, information (in this case, information about value) could potentially gain a selective advantage from the system in which it was represented, or from the medium in which it was realized. Languages coevolve with their media, and therefore the content and structure of each monetary system would have been shaped by the given medium of exchange. Conversely, each medium of exchange depended for its monetary function on a method of measuring its value. What is fascinating from a cultural evolutionary point of view is the way in which the early evolution of monetary systems mimicked the early evolution of writing.

How did early traders measure the value of the commodities that they exchanged? Where metal was the medium of exchange, relative value could be measured in the weight of an assayed piece. Where grain was the medium of exchange, relative value could be measured by weight or volume. In both cases, the need to check the quantity and quality of the medium of exchange in every transaction was eventually overturned by the advent of the first monetary symbols. I focus here on the evolution of monetary symbols on coins, but a parallel story can be told for the evolution of marks on warehouse receipts.

Over time, the utility of the metal that it contained became less important than a coin's symbolic value as a medium and standard measure of exchange. There are intriguing parallels between this process and the move from three-dimensional clay tokens to two-dimensional representations of goods for accounting purposes in Mesopotamia in the fourth millennium BC. Recall that in the first step beyond the use of individual clay tokens, several tokens were sealed within hollow clay envelopes, which would have to be unsealed to ascertain their contents. Comparably, the early use of precious metals as a commodity involved metal pieces that had to be assayed and weighed each time they were traded. For convenience, therefore, the Mesopotamian accountants began to mark the outside of each clay envelope with impressions of the clay tokens that it contained – just as the Lydian traders began to stamp their pieces of metal with marks that indicated the value of the metal they contained.

Finally, both the envelopes' contents and the actual value of the pieces of metal became obsolete: marks made with a stylus on a clay tablet could serve the same purpose as a hollow envelope with both token impressions on the outside and tokens on the inside, and marks on a cheap piece of metal could serve the same purpose as a piece of metal with both marks and an intrinsic value. In both cases, the significance of the marks lay in the changes in their representational content. Unlike the earlier token impressions on clay envelopes, the stylus marks on clay tablets did not represent the tokens whose shape they mimicked, but stood for the objects that those tokens themselves had represented: grain, oil and so on. Unlike the earlier stamps on precious coins, the stamps on the cheaper pieces of metal did not indicate the value of the metal they contained, but represented a promise of the payment of that value. In other words, the marks were a guarantee that the bearer of the coin could use it to purchase goods or services, or to settle debts, or to exchange it for precious metal, to the value that the marks represented.

Just as the original purpose of the marks on the outside of clay envelopes had been to represent tokens, which were themselves representational in nature, so the original purpose of the marks on metal coins was directly to represent the value of a commodity, which itself represented a particular amount of buying power. In both cases, what we can see is the evolution of a new system of representation, without which the development of neither a medium of exchange nor a method of accounting would have been possible.

As in other cultural areas, too, we can see the web of influence between the emerging artefactual languages and the media in which they were based. The medium in which it is stored can either facilitate or curtail the potential effects of information, as well as affecting the accuracy or detail with which the information can be stored. So, for example, the marks made on the first clay tablets were identical with the marks that had been made on hollow clay envelopes containing tokens, but the change of medium was massively significant for evolution in the content of those marks. Similarly, the shift from commodity money to credit money to fiat money was intimately related to the move from using valuable to near-worthless metal as the medium for coins. It is interesting to note, in this context, that as inflation detracts from the value that is measured by money, this is sometimes reflected in a medium change from paper money to coins. In other words, both the way in which money can operate as a medium of exchange and the way in which it can measure value will be shaped by the very medium that is used. If the medium is cattle, for example, then the high value that is measured by each unit of exchange will dictate the size of the exchanges that can be made.

Conversely, the medium's use as a means of exchange is dependent on its utility as a measure of value. First, items cannot be used as a medium

of exchange by people who have not agreed on their value, because
nobody will accept them in exchange for something that they do value.
This is obviously true of commodities (like grain), but it is also the case
for credit money (the buying power of which would disappear if the
commodities on which it is based were devalued) and for fiat money
(which is vulnerable to a loss of trust in the government or economy).
Secondly, being of intrinsic value within a society is a necessary but not a
sufficient quality for something to operate as money. A medium's intrinsic
value to both parties gives it the potential to be used as a means of
exchange, but so long as it remains only one among many candidates the
exchange is not monetary but barter. Monetary exchange, as Lapavitsas
has emphasised, is qualitatively different from barter in that it involves
the *universal* request to use money as the means of exchange. A medium
will acquire this kind of monopoly on buying power only if it lends
itself to use as a measure of value. For this, it needs to be the sort of
medium in which information can be discretely represented: it must be
durable and portable enough for the exchange to remain of value to
the recipient, divisible enough to be used in transactions of varying sizes,
and measurable enough for those divisions to be trusted as an accurate
indication of how much each portion was worth.

The better a medium's potential as a measure of value, the greater are
its chances of emerging as the cultural monopolist on buying power, repa-
ration and debt settlement. The medium will shape the ways in which the
information can be represented (whether as number of cattle, for example, or in a more sophisticated system), and in the competition between
units of cultural information, information represented in a system that
measures value in durable, appropriately sized chunks will have a replicative advantage over information represented in a system that discretizes
value into chunks too large to be useful; over information realized in
a medium that perishes before it can be used again for exchange; and
over information represented in a system that sellers do not understand.
To put this another way, representation was the driving force behind
the evolution of money as a universal medium of exchange. Of all the
things that society valued, it was the media in which value could best be
measured that came to dominate as media of exchange.

Different Forms of Money

The significance of the distinction between medium and language is
laid bare in the realisation of the same currency in different media, for

here we see the emergence of *the representational system* as the monopolist of buying power, whatever its medium. Money is whatever is accepted, within a group of people, as the universal medium of exchanging value. A group's acceptance of commodity money is based on their agreement that units of a specific commodity can be used to represent the value they wish to exchange. A group's acceptance of credit money is based on their agreement that the symbols of a specific currency can be used to represent a promise to pay units of that specific commodity. A group's acceptance of fiat money is based on their agreement that the symbols of a specific currency can be used to represent the value they wish to exchange. Once a group has made any such agreement, it is the currency, and not a particular exchange medium, which has the monopoly of buying power within that group – and there is no reason why the currency must be realised in a single medium.

Modern currencies can be exchanged and accessed in a multitude of different forms, including notes and coins, cheques, credit and debit cards, direct debits, standing orders, Internet banking transfers, ATM transactions, and so on. It is, perhaps, tempting to regard electronically accessed numbers as mere accounting records, and electronic transactions as taking place in the absence of a medium of exchange. The description of currencies as artefactual languages is helpful in revealing the error behind this assumption, however, because we can observe in other areas of culture the ways in which electronic impulses can act as media of exchange. If I want to pass on some information to you in written English, for example, then I can realise the information in ink on paper, in which case the exchange will be made by the physical act of handing over a marked piece of paper; but I could also realise the information in electronic form and make the exchange by sending you an email. So long as you can read written English and have a computer with Internet access, you can receive information from me in this way. All languages rely on the right kind of receivers – and money is no exception. "It is individual perception and the use of money which defines money" (Singh 2000: 3.4). Where value is measured in coins and notes, exchanges will take place visibly, by the physical act of handing over coinage, and this relies on the existence of people who have agreed to measure and exchange value in this way. Where value is measured in electronic banking records, exchanges will take place by the alteration of numbers in those records – and, for people who have agreed to measure and exchange value in this way, these electronic record changes provide the medium of exchange as well as a measurement of value. The mistake is to think that the electronic numbers represent a pile of notes and coins, so that alterations in the numbers are mere records of monetary exchanges yet to be made, or somehow constitute exchanges that are made in the absence of any monetary medium. In reality, the electronic numbers are representations of value, just as the coins and notes are, and value can be exchanged in whichever medium it is represented at the time.

Currency languages represent information about the amount of money that is owned or owed at a particular time, or received during a particular transaction. Like many other artefactual languages, currencies can be realised in more than one medium – and different media, as we have seen, possess different capacities for the preservation and transmission of information. The selection of a particular form of money for a particular transaction will therefore be influenced by the quantity and permanence of the information that users require from the transaction (Singh 2000: 3.16). When money is exchanged in the form of cash, for example, both parties to the transaction receive immediate information about how much money they had at the beginning, how much has changed hands and how much they have left: the information is represented on the coins and notes themselves, and these physically change hands, taking the information with them. For this reason, cash is a good method of transmitting information about value within a single transaction, but it is less adequate as a means of preserving information about multiple transactions, ownership and debt. This is why cash is avoided or supplemented in transactions where a more persistent record is required. When business expenses are paid in cash, for example, the buyer will usually request a written receipt: information about the value that has been exchanged is copied and preserved in a medium that is separate from the medium in which it was exchanged. Credit card payments, too, provide both printed receipts for individual transactions and monthly statements of the resulting debt. These pieces of paper, or electronic records, preserve information about the money transferred among credit card companies, buyers and sellers in a more permanent form than a cash transaction can manage.

We should not be surprised to find that currencies, like any other artefactual language, have coevolved with their media in particular cultural contexts, under selective representational pressure. In situations where one medium of exchange is not able to preserve value information with the persistency, accuracy or detail that its users demand in that context, value information is likely to be realised in a different medium. There is also evidence (Singh and Slegers 1997) that people are likely to restrict their monetary transactions to media that they trust – and indeed, as we have seen, all forms of money rely on their users' trust in the underlying financial system. When exchanges are made in media like cows and grain, traders (and other users of the monetary system) need to trust one another to continue to invest a consistent amount of value in those commodities. When money is exchanged in the form of credit notes, participants need to trust one another to redeem the notes. When money is exchanged in the coinage of fiat currencies, its users need to trust the government to maintain the value of those notes and coins. When money is exchanged electronically, its users need not only to trust the government to maintain the currency's value but also to trust the banks to maintain the records correctly.

Here again we can see the representational pressures on the artefactual languages of money, which propel their evolution in the directions of best fit to different cultural contexts. Cash exchanges of fiat money have less representational capacity and permanence than electronic exchanges of the same currency, but a greater speed of transmission; they are also restricted to the circle of people who physically own (or wish to own) that form of money, and for this reason, cash exchanges appear to many people to offer a guarantee of accuracy, which less tangible forms of money cannot provide. These factors fit cash exchanges best to circumstances in which users want to make an immediate exchange of value, in a form that they can physically hold and count for themselves. Electronic exchanges of fiat money have a greater and more persistent capacity for recording transactions than cash, but they are not always immediate, and they can be made with anyone who uses the electronic banking system. For this reason, money that is exchanged in the form of numbers, which are moved around in the course of electronic money management, is best fit for circumstances in which users want to make complex or frequent transactions, recorded in a persistent format, and they trust a network of third-party money managers to do this on their behalf.

Cash is used with decreasing frequency as a medium of currency exchange. More and more transactions are conducted in the various electronic forms that continue to evolve, and cultural evolutionary theory tells us that this is because the media of exchange are evolving to fit the different representational pressures that emerge with new types and increasing volumes of transaction: "the monetary medium can be vitally important to money's role in society, and to how people use and think about money in relation to their own lives and circumstances" (Dodd 2005: 564). At the same time, the languages of money continue to evolve, with the emergence of new forms like the euro to fit new relationships like those between the European nations.

In the unorthodox period between the euro's launch as legal tender in 1999 and the introduction of its coinage in 2002, the euro was traded on the financial markets and could even be used to denominate mortgages and loans within the euro zone, but it relied on the monetary media of individual member states for its operation as a medium of exchange. In practical terms, the absence of its own monetary medium meant that "the euro was not properly 'money' at this time" (Dodd 2005: 566). Even after the introduction of euro notes and coins in 2002, Dodd argues that the euro continues to operate in some senses as a "hybrid currency": a dual pricing system persists throughout the euro zone, and the designs on euro coins continue to include national symbols. Indeed, Dodd (2005) argues that the homogenization of currency in the euro zone may actually be stimulating the diversification of monetary forms. Dodd's analysis of the euro offers a practical demonstration of my theory that each type of money can helpfully be understood as the product of an evolutionary interaction between an artefactual language, its media and its receivers.

Money: Value, Exchange and Trust

Artefactual languages provide access not only to a particular portion of cultural information but also to a group of people who use the same artefactual language. Historical and current economic evidence supports a representational account of money as an artefactual language that represents information about relative value, and it suggests that the capacity to discretize value is what has given money its role as a medium of exchange. Conversely, the media in which currencies are realized shape the ways in which they can discretize value, and the representational advantages of some media are what led to their selection as the basis of a currency. When humans wanted to do business (whether commercial or social) with strangers, they needed a method of representing the value of what was changing hands. Representational pressure drove the coevolution of the artefactual languages (or currencies) that give these representations their meaning and the media in which they were realized. As artefactual languages, money's currencies provide functional links between the groups of humans who use them.

All forms of money have their basis in a collective agreement to measure value in a particular way and to exchange and realize its ownership in a particular medium. Barter is not money, because it is a not a representational system but a noncompositional series of bilateral value comparisons; an agreed system for the representation of value is not needed between people who can trust each other to make a fair exchange on any given occasion. Trust in a monetary system will sometimes be the result of its users' trust in an issuing authority, but in the view of money that I have developed in this chapter, the restriction of money to systems issued and regulated by institutionalized banking and political authorities (Ingham 1996:523) is arbitrary. All forms of money rely on differently sized groups to vouch for their acceptance (Simmel 1978: 178), but to claim that the group must include a fiscal authority, if its value measurements are to qualify as money, is simply to define all nonissued monetary forms out of existence.

A representational account of money is supported by Lapavitsas's (2003) argument that money is the facilitator of functional relationships between the social foreigners who adopt that system of measuring and exchanging value; but it suggests that a commodity's use as a measure of value will be the cause, rather than the effect, of its universal desirability and consequent monopoly on buying power. This view of money is also supported by Ingham's (1996, 2008) arguments that money is essentially a measurement of value, but it puts more emphasis than he would on the necessity for an underlying medium of exchange: all languages must be realized in a medium, even if the relationship between languages and media is coevolutionary and not always one-to-one. The truly distinctive feature of money is that it is an artefactual language, which is used by a given group of people for the representation of value, and is exchanged and realized in one or more representational media.

10

Money

The Explanatory Power of Artefactual Languages

Culturally acquired competence in an artefactual language provides access not only to the content that is represented in that language but also to functional cooperation with other people who are competent in its use. Such access, or conversely, exclusion, can have implications for individuals' social status and hence for their biological fitness. In this chapter, I show how a representational account of money has several explanatory advantages over alternative views of money and is also supported – rather surprisingly – by evidence from the Eurovision Song Contest. A representational view of money enables us to understand the connections between health, wealth and other measures of social status. By providing or preventing access to cooperation with people from social groups other than our own, culturally evolved artefactual languages can have an impact on our biological fitness – but this should not mislead us into thinking that they are therefore the product of biological evolution. The theory of evolution has revolutionised and unified our understanding of the natural world, and it can do the same for human culture – but only if we are careful to focus on evolution in the appropriate realm. The advantage of a representational theory of money is that it draws monetary phenomena under the same explanatory umbrella as every other aspect of culture: as the product of evolved cultural information, discretely represented in evolved cultural languages, which are realized in evolved cultural media.

The Effects of Currency on Trade

The view of money as an artefactual language, facilitating functional links between people who have no social relationships, predicts that currency should have a noticeable impact on trade. If the currency is doing its job as it evolved to do it, then communities that use the same currency as each other should trade with each other more easily than communities that use different currencies. The process is analogous to the process by which communities that share a natural language form social ties with each other more easily than communities that speak different languages: a shared language both facilitates socialising between those who speak it and excludes from the social circle those who do not. Indeed, some studies have indicated that "language differences may reduce genetic exchange between populations," and that as a result, "phylogenetic trees of major language groups correspond closely to phylogenetic trees constructed from human genetic markers" (Pagel and Mace 2004: 276). Similarly, we should expect to find that a shared currency will both facilitate functional (in this case, trading) links between those who use it and exclude from the trading circle those who do not. As always, heritable information depends on the shared use of a representational system: traders who do not measure value in the representational system of a particular currency cannot receive information about the value of goods when it is represented in that currency.

This result is just what the empirical data reveal (see, e.g., Tenreyro and Barro 2003). Andrew Rose (2000), among others, has done substantial research into this issue, and he concludes that one of the few undisputed gains from a currency union is an increase in trade among members. His results suggest that the impact is even greater than might be expected: countries within currency unions trade 300 percent more than nonmembers (Frankel and Rose 2002; Rose 2000). This "statistically significant and economically large" (Rose 2000) effect of currency unions cannot be attributed solely to factors other than the shared currency. Of course "economically larger and richer countries trade more; more distant countries trade less. A common language, land border and membership in a regional trade agreement encourage trade, as does a common colonial history" (Glick and Rose 2002). All of these factors might well be expected to increase both the chances that two countries will form a currency union and the amount of trade that they would do with each other, even in the absence of a shared currency. But even if we move away from "the cross-sectional question 'How much more do countries within currency unions trade than non-members?'" and consider, instead, "the (time series) question 'What is the trade effect of a country joining (or leaving) a currency union?'" (Glick and Rose 2002) the answer is still "substantial." In work with Reuven Glick (Glick and Rose 2002), Rose found that "a pair of countries which joined/left a currency union experienced a near doubling/halving of bilateral trade." This "economically large, statistically significant" result lends empirical support to a representational account of money.

Money and Biology

"Analyzing money as a cultural phenomenon, beyond immediate survival concerns, does not preclude tracing it back to its 'biological basis'" (Jorion 2006: 187). Our capacity for culture has its roots in our biological evolution, and it is no doubt the case that money's cultural evolution has been directed and accelerated by the biologically adaptive advantages, for humans living in social groups, of being able to exchange value with members of other social groups whom they encounter, whether in trade or in any other social context. Money's role as a measure of value helps to explain its undoubted adaptive advantages – both for communities whose increasing ability to measure value enabled them to use money as a means of exchange with other communities, and differentially for individuals within money-using societies. Indeed, Lea and Webley (2006) argue that trade can be seen as a form of reciprocal altruism,[1] because it is a reciprocal exchange that brings adaptive advantages to its participants, such as the availability of otherwise-unattainable goods and services, and functional cooperation with strangers.

The relatively recent innovation of money as the universal medium of exchange, however, has provided precious little time for our genes to extend their reach into our monetary transactions. The answer to the Dennettian "Cui bono?" might be "our genes," but a beneficiary is not necessarily identical with an executor. Although the strength of the desire for money (at least in some situations) makes it unlikely that there is no biological basis at all for that desire, the evidence shows that it does not act as an equally powerful motivator in all situations. I shall argue that when it does, this is because it is being used as a measure of social status.

My claim has been that the cultural evolution of artefactual languages has given humans the capacity for functional cooperation with a much larger group of people than would otherwise be possible. Money has culturally evolved as a system for the measurement and exchange of value, and there is an anthropological consensus that it has evolved for use in social as well as commercial contexts. We shall see in this chapter that,

---

[1] Reciprocal altruism occurs when organisms behave altruistically towards each other in the expectation of being repaid at some point in the future.

in some social contexts, the value that money measures is identified with social status, which is one of the strongest motivators for nonhuman as well as human primates. Nonetheless, it is equally important to remember that in some cultural contexts the question whether there is a biological reason why money is such a powerful incentive (Lea and Webley 2006: 163) is rather like asking whether a man has stopped beating his wife: like any other product of cultural evolution, money is not as successful in some cultural contexts as it is in others, and there are times when it is not a very powerful incentive at all.

Money Size Illusion

Some of the ways in which we interact with money are, of course, shaped by innate psychological mechanisms. It has long been agreed, for example, that we tend to overestimate the size of valuable objects, including money (Bruner and Goodman 1947). In particular, rapid inflation leads us to overestimate the size of obsolete notes and coins, and to underestimate the size of their replacements (see, e.g., Furnham 1983; Lea 1981). One explanation that has been offered for our skewed perception of currency size is that money acquires a special drug-like status from its value, which interferes with our normal processes of perception and cognition (Lea and Webley 2006: 169).

If we accept the evidence that our overestimation of the size of valuable items is due to an evolved psychological mechanism, then we must accept that our perception of notes' and coins' size will be affected by this innate aspect of our psychology: the extent to which we value money's tokens will affect our perception of their size. Indeed, the fact that an evolved psychological mechanism is implicated in a perceptual illusion is really not very surprising. But it is worth noticing that this explanation of money size illusion is incomplete. It explains our misperception of notes' and coins' size in terms of our perception of their value, but it does not tell us what affects our perception of their value.

The benefit of a cultural account of currency perception is that it can explain the money size illusion in the same way that it explains every other piece of monetary evidence: as the product of the usual processes of cultural evolution. On this account, currencies are representational systems, and individual coins or notes are representations of value. No receiver can acquire information from a source without first understanding the way in which that source represents information, because it is the overarching representational system which gives individual representations their meaning. Our perception of individual coins' and notes' value is therefore bound to be affected by our perception of their overarching currency. On this account, the money size illusion is the result of our changing perception of individual coins' and notes' value, which is itself the result of our shifting levels of confidence in the currency of which they are part.

Empirical support for this explanation comes from Leiser and Izak's (1987) work on the money size illusion in the context of high inflation and currency change in 1980s Israel. Their research effectively separated money's perceived representation of value from its actual market value, because they found that coins from the same obsolete currency were perceived differently, depending on whether they were still in circulation at the time that currency went into inflationary freefall. The actual value of all the coins in the currency was the same, relative to one another, so Lea and Webley's thesis would predict that they should all suffer from the same size illusion. In fact, though, coins that went out of circulation before inflation eroded the currency's purchase value were found to be immune to the predicted size underestimations. In other words, Leiser and Izak found that the underestimation of coinage size was not caused by inflation itself (i.e., a change in the actual value of coins and notes). "Rather, inflation causes a loss of confidence in the currency, and this climate is reflected in the size estimations" (Leiser and Izak 1987: 355). They concluded "that the main factor operative in the money size illusion is the attitude of the public towards a given coin" (1987: 354): sliding confidence in a currency produces a decrease in its perceived value and a consequent decrease in its perceived size. This research effectively undermines Lea and Webley's hypothesis that the money size illusion is created by changes in the actual value of individual coins or notes, and supports the representational view that our perception of the value of individual coins and notes will be affected by our perception of the way in which value is represented by the currency of which they are part.

Leiser and Izak's results illuminate another interesting facet of our dealings with money: we are not very good at keeping up with changes in its value. Coins from an obsolete currency retained public confidence so long as they went out of circulation before inflation eroded that currency's value. Clearly, this undermines Lea and Webley's contention that the value of a coin determines its perceived size, but doesn't it also contradict my suggestion that a coin's perceived value is affected by its currency's perceived value? The public was confident in these coins, even when they had lost confidence in the currency of which they used to be part. The key phase, here, however, is "used to be": those coins were never a part of a discredited currency, because they had gone out of circulation before confidence was lost. Public perception of their value was never updated to match the sharp decline in their actual value.

Here, too, we can see another crucial cultural factor at play: it is always *perceived* value that determines cultural success; evolution is not interested in absolute truths but only in relative fitness. Our inability to keep up with monetary changes is also seen in the persistent illusion, after a period of rapid inflation, that money can buy more than it really can. "Money illusion disconnects the psychological impact of money from what money can do" (Lea and Webley 2006: 169), and this is precisely what a representational view of cultural evolution theory predicts will happen throughout human culture.

Money in Society, Money in Relationships

Other aspects of our interactions with money are dictated more by cultural norms than they are by biological psychology. The use of money cannot be disentangled from its relational context (Zelizer 2005): we even have different words for different ways of handing over money – payment, gift, salary, tip, repayment, bribe, compensation, and so on – and we go to great lengths to ensure that our economic practices conform to our understandings of our relationships. In modern Britain, for example, it is not often acceptable to repay neighbourly help with money, and there are fairly clear social rules about when money is or is not acceptable as a gift, depending on the ages, relative status and relationships of the donor and recipient (as discussed later in this chapter). More generally, there is a cross-cultural tendency to distinguish between market exchanges, in which money is not only acceptable but also usually required, and gift exchanges, in which money may or may not be acceptable (Lea and Webley 2006: 170). Within sexual and marital relationships this distinction is particularly applicable, but Lea and Webley discuss several other situations to which it is also relevant, including the revulsion that most people would feel if asked to estimate the market value of their children, friendship, and so on. In these ways, money's role as a medium of exchange is less universal than we might at first assume. The point is not that circumstances exist in which money *cannot* be used as the medium of exchange, but rather that there are circumstances in which it is not socially acceptable to use money as the medium of exchange (Lea and Webley 2006: 171).

From a representational point of view, such restrictions on money use are unsurprising. Because money functions as an artefactual language that enables people to do business with each other in the absence of social ties, we should not expect it to be seen as an appropriate medium of exchange between people within an existing relationship. If money is the product of cultural evolution for a medium of exchange between Lapavitsas's foreigners, then its attempted use within relationships is bound to be insulting.

Further evidence for the distinction between social relationships that are built with the help of communicative languages, and functional relationships that are built with the help of representational languages, comes from the way in which advertising and sales techniques exploit the nature of our reaction to the personal approach, disguising commercial transactions as reciprocal personal relationships. "In the web of commerce, every human bond is open to betrayal," because where money is involved, only "*pseudo regard*" is available (Offer 1997: 467). The aim in "relationship selling," for example, is to tap into the reciprocity of personal relationships, creating an obligation to buy because we feel that we have received some token of regard from the salesman (Offer 1997: 466). Perhaps someone we know has invited us to a party or other event: "Tupperware, Avon, and Ann Summers recruit women to draw on their social networks and convene house parties, where the conventions of female reciprocity are invoked to sell plastic tableware, cosmetics, or sexual accessories" (Offer 1997: 465). Business lunches perform the same function.

This distinction between social and functional relationships, which sales techniques seek to blur, is all that we need to explain the restrictions on money use. There is a range of evidence that people's behaviour towards one another is affected by the extent to which they have been prompted to think in monetary terms at the time (Vohs, Mead and Goode 2006, 2008; Zhou, Vohs and Baumeister 2009). For instance, Vohs, Mead and Goode (2008: 211) report unpublished findings from their laboratory, which suggest that "after people are reminded of money, they show improved memory of exchange-related information, prefer exchange-based relationships, and follow equity rules." The authors also cite research (Kasser and Ryan 1993) which indicates that "Americans who highly value money have poorer relationships than do those who take a more moderate approach to money."

We do not offer money to our neighbours for their help, because this is not an exchange between strangers. We do not offer money for

sex unless it *is* an exchange between strangers. We should feel peculiarly affronted by a request to estimate our children's monetary value,
because we love our children. The financial basis of a marriage is made
explicit in the process of divorce, in a way that would not be contemplated
within an existing marriage, precisely because the relationship has broken down and the two people have once again become strangers. There
is evidence that financial issues reliably predict divorce (Lea and Webley
2006: 171), and according to the representational view of money, this is
because the lasting importance of money issues is inversely proportional
to the strength of a relationship: in other words, money issues predict
the end of a relationship because they reveal its disintegration. Money is
"a potent symbol and channel of the power relationships within a family"
(Lea and Webley 2006: 171) only when family relationships are under
strain. Where relationships are solid, money is not an issue. Where they
are weak, individuals become conscious of any financial imbalance. The
"division of labor, both domestic and paid, has been shown to be a weak
predictor of marital satisfaction for both husbands and wives (Wilkie,
Ferree & Ratcliff, 1998): researchers found that this relationship was,
instead, largely mediated by how equitable each partner found the division to be, as well as the amount of empathy each spouse had for the
other" (Brockwood 2007). In other words, spousal attitudes towards the
financial balance within a marriage are dictated by the state of the spousal
relationship and affected by preexisting expectations of what a marital
relationship should be like (Booth 1979; Perry-Jenkins and Folk 1994).
They are not dictated by absolute facts about who does and who does not
earn or have access to money.

Money as a Gift

In particular, a representational view of money can account for the social
rules, which vary across cultures, about when money is an acceptable
gift. Why should its acceptability depend on the relative age and status
of the donor and recipient, and why shouldn't we evaluate a gift purely
financially (Lea and Webley 2006: 170)?

Within the context of relationships, gifts are given as tokens of what
Avner Offer (1997: 451) calls regard, an attitude that "can take many
forms: acknowledgement, attention, acceptance, respect, reputation, status, power, intimacy, love, friendship, kinship, sociability." For this reason,
Offer argues that money will not be viewed as an acceptable gift in the
context of many relationships, because it is perceived as impersonal. "A
gift, on the other hand, is *personalized.* Even when obtained from the

market, it provides evidence of an effort to gratify a *particular* individual" (Offer 1997: 454). For this reason, cash is acceptable as a gift only if it is in some way "transformed" (Singh and Slegers 1997: 17), such as by gift wrapping it or, in that most bizarre of gift-giving practices, by first exchanging a quantity of money, which can be used anywhere and purchase anything, for a voucher that can be used only in a particular shop. We do not evaluate a gift purely on its monetary value, because gifts are media of emotional rather than merely of financial exchange.

It is important to note, in this context, that there are varying social rules not only about when money can be used as a gift but also about what sort of nonmonetary gifts it is appropriate to give. Because a gift is something that is given within the context of a relationship, it is the nature of the relationship that will determine what is appropriate as a gift. Sometimes, the nature of the gift can be ill judged, inaccurately reflecting the nature of the relationship, and in these cases it can give offence rather than pleasure. Within a new sexual relationship, an overly expensive gift might be misinterpreted as an attempt at payment; within sibling relationships, an imbalance in the value of gifts given can be seen as inappropriately competitive or patronisingly charitable. It is easy to imagine a whole range of situations in which a chosen gift might embarrass the recipient, its inherent intimacy or generosity not reflecting the true nature of that relationship.

Conversely, there are times when money can be used as a gift, because the transfer of money between donor and recipient does accurately reflect the nature of their relationship. If a gift is meant to reflect the relationship between donor and recipient, then there are situations in which a relationship is such that monetary gifts are gladly received: in modern British culture, for instance, few children would take offence if they found a banknote in their birthday card from an uncle, god-mother or grandparent. Money is appropriate in these situations for three reasons. First, it is acceptable for a gift to be impersonal when the relationship between donor and recipient is actually rather impersonal: a gift may be given within the context of a formal relationship within the family or friendship circle (e.g., aunt to niece), even when the older donor does not actually know the younger recipient very well. Secondly, a gift of money will accurately reflect the nature of many relationships between older and younger generations, particularly within the family but also including godparents and other family friends: the older generation provides for the younger generation. Thirdly, and relatedly, money is acceptable in this situation because it ends the recipient's obligation to reciprocate.

When we pay for something, the transfer of money ends our relation-ship with the vendor: she has provided goods or services, we have paid for them, and our payment marks the end of that functional relation-ship. This is one reason why, normally, money is not an acceptable gift: whereas gift exchange is an ongoing, reciprocal process of turn taking that reflects an ongoing, reciprocal and equally balanced relationship, money has evolved to facilitate one-off transactions between foreigners. Money does not work as a gift, because it would make no sense for there to be an ongoing, equally balanced process in which the participants take it in turns to give each other similar amounts of money. Indeed, it can be insulting as a gift, because the implication is not only that the donor takes the role of provider in the recipient's life but also that the process of exchange between them is at an end. Only when the donor actually is a provider can money to be used as a gift, because in these circumstances both parties know that there is no obligation to reciprocate, so the end of this particular exchange need not mark the end of their personal relationship.

The evidence is, on balance, that there are restrictions on the use of money as a gift because we are able to distinguish between social and functional relationships. Gifts are meant to reflect the nature of the rela-tionship between donor and recipient, whereas money's cultural evolu-tion was dictated by the need for a medium of exchange that would work outside personal relationships. Money is precisely that which does not reflect any ongoing, personal relationship between its users; it is use-ful as a medium of value exchange but not as a medium of emotional exchange. "Real regard is typically not for sale" (Offer 1997: 454).

Eurovision: A Diversion

Further evidence for a representational view of money comes from a most unlikely quarter: the Eurovision Song Contest. In short, the evidence reveals that, even when a different system is used for the exchange and representation of value, a very similar pattern of exchange, restrictions and tensions emerges. In other words, the representational account of money is strengthened by the fact that it can also explain what is going on in an entirely different system of exchange.

The Eurovision Song Contest is an annual competition in which mem-ber states of the European Broadcasting Union enter songs, of less than three minutes' duration, which have not previously been commercially released. When all of the songs have been performed, each country ranks each of them and assigns it a certain number of points, from twelve for the best-liked entry down to one point for the tenth best-liked entry, awarding zero points to any remaining countries. Countries cannot vote for themselves, and the winner is the country with the highest number of points.

There has been much analysis of Eurovision results, and particularly of the results in the years since 1975, when the current scoring system was introduced. Although the factors affecting some countries' voting patterns are impenetrable, research indicates "strong evidence for voting bias in the Eurovision Song Contest, based on geographical, cultural, linguistic, religious, and ethnical factors" (Spierdijk and Vellekoop 2009: 424). The subjective factor in musical evaluation means that a high estimation of foreign songs depends "on a cultural match between the evaluator and the evaluated" (Yair 1995: 149), and a social network analysis of Eurovision voting patterns from 1975 to 1992 reveals a voting matrix that reflects Europe's underlying cultural and political structure (Yair 1995; Yair and Maman 1996). More specifically, statistical analysis reveals that linguistic and cultural similarities are a better explanation of voting patterns than, for example, political alliances and rivalries between countries (Ginsburgh and Noury 2008), and that "the general public exhibits these biases in their voting pattern in stronger terms than juries of experts" (Clerides and Stengos, 2006: 26).

The fact that countries award more points, on average, to songs in a related language to their own, which come from countries that are closer in terms of geography, culture and religion to their own (Spierdijk and Vellekoop 2009) should, it might seem, be welcome news for a representational view of cultural evolution. A shared musical heritage affecting judges' subjective responses to songs? That's just what a representational analysis of music would predict (see Chapter 8). A shared language increasing the trade in votes? It sounds remarkably similar to what happens when a shared currency increases trade in commerce. As is so often the case in cultural analysis, however, the situation is more complex than such a simple comparison might suggest. A shared currency is not, in any case, the only factor that influences the amount of trade between two countries: what is important about Rose's research (see, e.g., Glick and Rose 2002) is that it exposes currency as a factor *in addition to* the more predictable indicators like cultural and linguistic similarities. It reveals that currency, which is an artefactual language, creates functional links between countries, even in the absence of social links.

An important difference between these two sorts of links, as noted in the previous section, is that the exchange of money for goods or services draws a line under the transaction, whereas social ties give rise to ongoing, reciprocal exchanges between individuals (or even nations). Derek Gatherer's (2006a: 1.12) "rigorous analysis of changes in collusive voting patterns over the history of the contest" suggests that the same distinction is at play in Eurovision. Gatherer's (2006a: 4.4) research reveals a "progressive increase since the mid-90s" in the "non-randomness and suggestive internal structure in the voting patterns," and argues that this rapid growth shows that cultural preferences cannot be the only explanation of Eurovision voting blocs. "In the 1980s, there were only two or three countries that were involved in observable vote trading partnerships. But from the 1990s onwards it increased dramatically. In 1993, there were six countries involved. In 1998, there were 12 countries, and now we have 31 countries involved" (Derek Gatherer, quoted in Alexander 2008). It is likely, Gatherer (2006a: 4.8) says, that "collusive voting has increased because voters have realized that it increases their own country's chance of winning the contest."

At first glance, there would appear to be a degree of tension between Gatherer's account of Eurovision voting patterns and the alternative suggestion that they correspond to preexisting patterns of linguistic and cultural similarities. As Gatherer points out, linguistic and cultural factors cannot be the only key, here, because they do not explain why voting patterns should have altered so dramatically over the years. But this conflict is defused as soon as we take into account artefactual as well as natural languages, and the related distinction between functional and social ties.

The Eurovision voting system is as much an artefactual language as money is: the number of points that one country gives another is a representation of the value, or quality, of that country's song. Twelve points represent the highest possible quality: better than all of the other songs in the competition that year. Zero points, conversely, represent the lowest possible quality. Within the limited context of Eurovision, the voting system also facilitates exchange. Just as we offer an appropriate amount of money, in the marketplace, for the value of the goods or services on offer, so in Eurovision participants offer the appropriate amount of points for the quality of the songs on offer. Like a monetary exchange, the exchange of points for quality both facilitates the functional interaction between participating countries (in other words, without a voting system, the competition could not take place) and draws a line under the transactions between participants. One country has offered a song; the other countries have adjudged its value – end of story.

Except that what the statisticians have revealed is that, increasingly, this is neither the end nor indeed the beginning of the story. The problem, from a functional perspective, is that participating countries have an established pattern of interrelationships, and ongoing patterns of reciprocal exchange. This has always been the case, of course, but in 1998 a substantial innovation was made in the system of allocating points to each song: jury voting was replaced by "televoting." Under the old regime, a small panel of expert judges in each country had assigned points to every entry. Every judge did of course have a nationality, but in playing the role of a Eurovision judge he was asked to put this aside. In effect, this was a functional transaction among strangers: the voting system was designed for the exchange of value between singers and juries, for the purpose of facilitating Eurovision. In the background, meanwhile, were millions of viewers throughout the participating nations and beyond, almost all of whom were inevitably partisan. So we can see that there were two parallel processes at this stage: a one-off, functional exchange of value between songs and jurors during the competition; and an ongoing social exchange of regard between competing nations.

What happened in 1998, when the voting was taken out of the hands of a small number of experts and given to the wider public, was that the barriers between the two processes dissolved. A system that was designed for the exchange of value between singers and jurors was hijacked for the exchange of regard between nations. National preferences had always existed, but the juries' relative independence had insulated the results of the competition from them. Under the new system, it was as if responsibility for the outcome of a national football league had been passed from the competing teams to their fans: the results were being shaped not by the quality of individual performances, but by what the fans wanted to be the case. Like a football game, a song competition is meant to be a functional, one-off exchange between strangers, and the rules of the competition are meant to facilitate this. If relative numbers of points are assigned not by an impartial method like sporting regulations or expert judgments, but by the votes of people who are entangled in an ongoing net of reciprocal relationships that both predate and postdate the competition, then it is inevitable that the outcome will reflect the pattern of those relationships rather than the genuine value of entries.

This explanation enables us to reconcile Gatherer's account of Eurovision with the view that voting patterns are skewed by culturo-linguistic relationships. As Gatherer (2006a: 4.8) points out, in the era of jury voting there were only a couple of pairs of persistent vote traders in the competition, but since the introduction of televoting there has been an increase in collusive voting, because voters quickly recognized the voting impact of the "long-standing Greek-Cypriot partnership, which probably did originate as an expression of political solidarity," and they copied it as a means of obtaining more votes for their own countries. The patterns of collusion that consequently emerged *can* (largely) be explained by the culturo-linguistic similarities between voting blocs of nations (Spierdijk and Vellekoop 2009), but so can the fact that collusion emerged at all (Gatherer 2006a). There are two parts to that explanation. The first is that the outcome had been taken out of the hands of strangers (the expert jurors) and given to groups of people who already had social ties with one another: this took the whole competition out of the context of a one-off, functional exchange and placed it firmly in the context of an existing pattern of reciprocal social ties. Patterns of collusion were bound to result within this context – and the second part of the explanation for their emergence is, as Gatherer points out, the most basic cultural evolutionary explanation of all: when people see a good thing, they copy it if they possibly can.

A representational view of money, then, as an artefactual language that provides functional links between strangers, is supported by evidence of what happens when a different system is used for the exchange and representation of value. In Eurovision, participants use points to measure songs' worth rather than currency to measure commodities' value, but the voting system is nonetheless a method of measuring and exchanging value in a functional context. In both monetary and voting transactions, the introduction of existing social relationships interferes with the proper measurement of value. When people care more about their social ties than they do about a fair exchange on this particular occasion, reciprocity replaces one-off exchanges and the system collapses.

Money and Status

We have seen how a representational account of money is consistent with evidence from a surprising range of disciplines – including archaeology and anthropology, economics, psychology and sociology – and how it can contribute to debates within many of those areas. Finally, I turn to what is perhaps the most persuasive evidence in its favour: its capacity to explain both the significant genetic benefit that money *can* bring to those who have enough of it, and the social situations in which this does not obtain.

Research indicates that, even in modern Western societies, a lack of money is associated with adaptive disadvantage in the sense of shorter life expectancy and poorer health. Wealth, as one study into twenty-first-century health inequalities in the Northwest of England put it, means health (Wood et al. 2006). According to this study, a range of conditions from heart disease to self-harm, alcohol-related deaths, diabetes, epilepsy and asthma showed a strong relationship with financial deprivation. Other studies have revealed that discrepancies continue to emerge even at the highest reaches of the income scale: at the turn of the century, Americans whose income was a substantial $100,000 a year nonetheless suffered more health problems and lived shorter lives than those who made $500,000 a year (Clay 2001). Why should this be?

One explanation of the relationship between relative wealth and health is that a person who makes $100,000 will be aware that there are others who make $500,000 a year, and this may have an impact on his self-perception, which could explain the inequality in health outcomes. Indeed, research into socioeconomic status as an aetiological factor indicates that it is an individual's *subjective* perception of his own social status that is most strongly related to a whole range health indicators, from mortality to depression, diabetes to cardiovascular risk, respiratory illness to self-rated health (Adler and Stewart 2007). This discovery grew out of an initial study of the link between stress and upper-respiratory infection in monkeys. "Although stress had no effect on the monkeys' susceptibility to the virus researchers exposed them to, it turned out that their position in the monkey hierarchy did. The more subordinate the monkey, the more likely it was to succumb to the virus" (Clay 2001: 78). Like the monkeys, humans who believe that they have a low social status have a disproportionately high risk of developing infections (Cohen 1999).

There is another similarity between the response of human and non-human primates to relative social status, as revealed in the effects of psychological stress on food preferences and consumption in socially housed adult female macaques (Wilson et al. 2008): socially subordinate females, whose levels of chronic psychological stress are measurably higher than those of their social superiors, will normally eat a little less than those superiors; but when a high-fat, sugary diet was made available to them, the lower-status females began to eat significantly more and gained weight as a result. In a different experiment, Morgan and colleagues (2002) showed that socially housed macaques were more likely to become addicted to cocaine, when it was made available to them intravenously if they pushed a lever, if they were of a lower social status. These results chime with evidence of the effects of psychological stress on food preferences and consumption in adult women (Epel et al. 2001; Zellner et al. 2006) and are consistent with emerging evidence of the association between obesity and poverty in humans, even in developing countries (James 2004). It seems that eating sweet, high-fat food can provide comfort to both human and nonhuman primates under psychological stress, and that relative social status is a significant cause of psychological stress.

In fact, self-reported status turns out to be more highly correlated with biological and psychological health outcomes than objective indicators like income and education. For instance, an objectively low socioeconomic status may be mitigated by a subjectively high community status, as when an out-of-work person living in a deprived area is nonetheless highly respected for her valuable voluntary work in the community. Does this mean that money is not, after all, adaptively advantageous? The emphasis on subjective assessments of status certainly cautions us against making simplistic associations between money and genetic advantage, but the link (if not equivalence) between health and status – and, relatedly, between status and wealth – is nonetheless compelling. What we need, as Paul Jorion (2006) has put it, is a three-term model of money, which acknowledges that individuals do not desire only money (a two-term relationship) but also that others should see that they have money. The fact that there are similar links between health and status in nonhuman primates provides further evidence that there would have been adaptive advantages to early humans who could increase their status by the judicious manipulation of wealth. "The overall benefit of admiration is fitness or reproductive advantage" (Jorion 2006: 188).

The Survival of the Joneses

The subjective factor in health outcomes provides further support for the thesis that it is money's use as a measure of value – its cultural evolution as the medium in which information about value is represented and exchanged – which explains its biological adaptiveness. For if people use a different measure of value, then money becomes less strongly related to physical and psychological well-being. The advantage of the representational view of money is that it can account for what goes on in these situations, too – just as an earlier section showed that it can also account for what goes on in Eurovision.

Our biological fitness is affected by our perceived social status, and liquidity is not the only factor in our self-perception: in a highly literate society, for example, the impact of illiteracy on self-perception can be devastating; within the academic community, great store is set by the number of occasions on which one's work is published and then cited by others; a footballer who plays in the top league will be more highly regarded by other footballers than one whose club competes in a lower league; and so on. These sorts of factors in our self-perception feed into humans' desire to fit in with their social group: to be valued highly, when judged by the standards of that group. As a British politician has observed, "we seem unable to judge ourselves, and to value ourselves, except by reference to other people. If we see others doing much better than us, we often feel threatened and unhappy; and we too often feel reassured to see someone else fail, or get their comeuppance" (Johnson 2007). In other words, we judge ourselves by the standards of our group, and our competition for status is played out in accordance with the group's rules.

In many social groups, money remains the accepted measure of value: people's self-perceived status is determined by their relative prosperity and its material expression in clothing and décor, cars and houses, holidays and other lifestyle factors. But human society is made up of all sorts of different groups, in which members might be linked by ties of kinship or geography, natural language or religion, shared interests or skills, and so on. In many of these groups, value is not measured by money, because it is not money that provides the links between members: rather, they have a shared competence in a different artefactual language – and for this reason, group members judge themselves against nonfinancial standards.

In the scientific community, for example, David Hull has persuasively characterized "the social structure of science as something akin to a market mechanism. Intellectual credibility, which Hull calls 'credit', takes the place of profit and empirical knowledge takes the place of economic product. The reward structures of scientific institutions are such that to achieve their individual goals, whatever these may be, scientists must accumulate credit and to do that they must contribute to the production of pragmatically effective empirical knowledge" (Griffiths 2000: 304). What Hull (1978: 685) has identified is the coincidence of "the selfish goals of individual scientists" with "the manifest goal of the institution, the increase of empirical knowledge" – a coincidence that is, itself, the product of evolutionary processes. Within the scientific community, intellectual credibility is exchanged for empirical knowledge, and status is determined by one's stock of intellectual credibility rather than by one's wealth.

Science is a particular example of a more general phenomenon that sees group members judging themselves against a variety of nonfinancial standards. Moreover, one person can be a member of any number of these different groups, by virtue not only of biological factors like her family of origin, sex, age and life stage but also of cultural factors like her culture of origin, natural language dialect, faith, education, hobbies and skills. Within each of these groups she is judged by a particular set of standards: she might have a very high status within her family and faith, for example, but be judged poorly according to material or academic standards. Her self-perceived status in this case will depend on which group she herself values most highly. With which group does she most identify? This question matters because there is often a tension between the standards of different groups. A significant part of the process of ado-lescence, for example, is a transfer of loyalty from family to peer group, which can involve considerable experimentation to discover the group to which an individual feels that she actually belongs (Distin 2006b: 179). For a bright teenager who still feels most at ease in her family of origin and who has little regard for her age peers' values of popularity, fashion and conformity, the social price that she pays for this attitude might have little impact on her self-perception. A different personality, whose good behaviour and examination success have brought her admiration and rewards from her parents and teachers, might nonetheless suffer from a crippling lack of self-esteem, if what she cares about at that time is the standards by which other teenagers will judge her.

At a less intense level than this adolescent self-torture, all humans can perceive themselves as members of more than one group at a time. In fission-fusion societies like ours, and like chimpanzees' and some other creatures', "members frequently coalesce to form a group (fusion), but composition of that group is in perpetual flux, with individuals frequently departing to be solitary or to form smaller groups (fission) for a variable time before returning to the main unit" (Barclay and Kurta 2007: 44). It is becoming apparent, however, that the degree of fission-fusion dynam-ics varies markedly, both between and within species (Aureli et al. 2008). In bonobo society, for example, large groups will sleep in one place but split into smaller groups for foraging. Within bat colonies, although members may roost together in one tree at any given time, there are always individuals who are roosting on their own or in small subgroups elsewhere, and the habits of each colony member will fluctuate over time (Barclay and Kurta 2007: 44). Individual bottlenose dolphins form small groups whose composition changes daily or even hourly (Connor et al. 2000: 91). In the light of such varying patterns, and as a result of recent collaboration between nineteen scientists from a variety of disciplines, agreement has been reached that the best use of the term *fission-fusion dynamics* is "to refer to the extent of variation in spatial cohesion and individual membership in a group over time" (Aureli et al. 2008). This fission-fusion structure has a clear impact on individuals' relationships with one another. "Unlike animals that live in groups of constant composition, social relationships in a fission-fusion society may depend strongly on the social context: who is there and who is not" (Connor et al. 2000: 91). Whereas the social challenges to members of more stable groups will focus around the need to avoid social disintegration by resolving the sorts of conflicts that arise from relatively high levels of association, members of groups with a higher fission-fusion dynamic must be able to find ways to resolve uncertainties and reestablish relationships between members whose association is periodically broken (Aureli et al. 2008).

Because human health is correlated with self-perceived status, our membership of these fluctuating groups is bound to affect biological fitness. If I perceive that my status is low in one group – if, for instance, I have sunk to the bottom of the academic pile at school – then my well-being can be improved if I gain access to a different group, where status is measured by different standards: a peer group that I can impress by my truculence and rule breaking, perhaps. This particular group might be accessed by forming social ties using natural language, but other groups will be accessible only by use of a shared artefactual language. By providing a way into different cultural groupings, money and other artefactual languages thus increase individuals' chances of finding an arena in which their self-perception will be positive, because they are being judged against standards that suit them. Artefactual languages, of which money in this context is emblematic, provide not only links between otherwise-unrelated groups of people but also the standards against which members of those groups can judge themselves. In this way, these cultural artefacts have an impact on our biological fitness, both by giving us functional and social access to a wider range of people and by providing alternative standards against which we can be judged. Although they are not biological in origin, therefore, the cultural evolution of artefactual languages may have been propelled by the biological advantages that they brought – just as the cultural evolution of natural language was propelled by the biological advantages that it brought.

A representational account of money has the virtue of bringing a range of monetary phenomena under the same explanatory umbrella as the rest of culture. The key to this explanation is the distinction between the social relationships that are supported by natural language and the functional relationships that are supported by artefactual languages like money. The connections between health, wealth and other measures of relative social status demonstrate that biological advantage can be gained as a result of the intergroup access that artefactual languages facilitate. Money is a system for the exchange of value between strangers. Its acquisition is a particularly powerful incentive in situations where it is used to measure not only commercial but also social value, because social status is one of the strongest motivators for both human and nonhuman primates. Its acquisition is a less powerful incentive in situations where its use is purely commercial and social status is measured in other ways, although its commercial utility in these cases will still have some effect on people's behaviour.

What begins to emerge from this account, therefore, is the range of human responses to our cultural inheritance. The next chapter explores one of the key aspects of human diversity, and its impact on cultural evolution.

PART IV

THE RECEIVERS OF
CULTURAL INFORMATION

11

How Does Human Diversity Affect
Cultural Evolution?

Cultural evolution is founded on the persistent inheritance of cultural information via the mechanisms of natural and artefactual languages. These fabulously intricate and capacious representational systems enable cultural resources to be preserved and their variations inherited. Information is any variation that a receiver discretely represents, and humans are uniquely able to learn discrete, compositional systems of representation from one another, which prepare us to receive more cultural information. We are innately disposed to acquire the local natural language as infants, but our metarepresentational abilities enable us to go on learning new languages throughout our lives.

Because information's inheritance is so dependent on the nature of the receivers involved, the nature of the receivers is bound to affect the patterns of its inheritance and, consequently, the course of its evolution. As has already been noted, humans have a unique and universal capacity for metarepresentation, which has been crucial for cultural evolution. What has not so far been acknowledged is the considerable variation in this universal human capacity. This chapter argues that individual metarepresentational ability has a genetic basis, because it is closely related to general intelligence, but that it can also be culturally enhanced, especially by education. In addition, it argues that although there is a fairly complicated relationship between *genetic* fitness and intelligence, individuals who fall at the higher end of the metarepresentational spectrum are important engines of *cultural* evolution. Humans are the agents of cultural evolution, and humans vary in ways that affect the transmission and expression of the cultural information that they receive.

The Metarepresentation Spectrum

Humans, as we have seen, are genetically prepared to be receivers of cultural information and the languages in which it is transmitted. Our metarepresentational capacity is an important facet of this genetic preparedness. Every human who is not severely mentally impaired is capable of metarepresentation, although individuals vary in the extent to which they engage in metarepresentational thought. This type of thought is essentially founded on pattern recognition: the ability to notice the similarities and differences in different portions of information and the ways in which they are represented. Evidence from studies of gifted children indicate that the tendency to think in this way varies among modern humans, and that a key quality of the very brightest individuals is their bias towards this way of thinking. Indeed, when you begin to look into the characteristics of people whom we might describe as gifted, what you find is that they are remarkably similar to the characteristics of a person who is exceptionally good at metarepresentation.

When we describe children or adults as gifted, we do not necessarily mean that they are geniuses, but rather that they are very bright sparks whose intelligence falls within the top 5 percent or so of the population. Intelligence is one of many human characteristics that, roughly speaking, follow a pattern called the normal curve of distribution, or the bell curve. What this means is that, like height and weight and longevity, intelligence is one of the areas in which most people are fairly average. Statistically speaking, there is not very much difference in the height, weight, life expectancy or intelligence of about two-thirds of any given population. Most people are pretty average in these areas: somewhat above or below, perhaps, but not very different from what we might think of as normal. The further away we get from the average, however, the more rarely do we see people of a particular height, weight or level of intelligence.

Children learn both natural and artefactual languages at varying ages and with varying levels of ease. There is significant individual variation in the vocabulary size that each person attains, and this variation is significantly correlated with genetic factors (Hurford 2007: 238). Intellectually gifted children are often noted for their early language skills, both in speech and in comprehension, and of course they are frequently very early readers, too (unless they are doubly exceptional, such as dyslexic as well as gifted). As they grow older, what distinguishes these children from their age peers is their enhanced ability to make connections among diverse subject areas; to see relationships and make generalizations from only a few given facts (Distin 2006b: 22–42). Whereas most children might need to work through forty examples before they can grasp a new mathematical concept, for example, a very bright child will understand it the first time that it is explained. Gifted children are capable of abstract thought, of humour, and of debate and logical argument at much earlier ages than other children are. They make original connexions, and their logical reasoning can lead them fearlessly to challenge orthodoxy: where the majority of people might accept and agree with what they are told, the gifted child is constantly popping up and shouting, "But the emperor has no clothes!" What distinguishes the very brightest children, in other words, is their unusual capacity for metarepresentation.

Metarepresentation and Cultural Transmission

Giftedness is not only an intellectual characteristic but also a quality that pervades the whole personality, in ways that affect other people's perception of a gifted individual from the moment of her birth. Crucially for cultural evolution, it also affects the ways in which she interacts with the cultural information that she receives.

Metarepresentation, the ability to make and learn connections, is what enables us to learn and use languages. We can learn how individual representations gain their meaning from a representational system, because we can see the connections that hold the system together. The system, or language, discretizes its content in a particular way, and we need to learn to discretize the source in the same way if we are to acquire information from it. Once we have learned the right pattern of connections, we can acquire information from any source that represents information using that representational system. As has been noted, this also means that the information we acquire from a representational source will be discretized in a particular way: the language in which it is represented shapes, to a certain extent, the conceptual framework of the information that we acquire.

The beauty of the metarepresentational capacity, however, is that it enables us not only to learn existing languages but also to create new representations and representational systems. Highly metarepresentational individuals are likely to see connections where others have not noticed them, by virtue of which they can create new cultural information. At the simplest level, if you have two existing representations and you notice a connection between them, then this makes you a receiver in the know about something new: you are now able to treat something as a source of information, when previously you were not. Not until we have connected our representations of domestic dogs and wolves, for example, can we use domestic dogs' behaviour as a source of information about their lupine ancestry. At a more complex level, metarepresentation enables us to shift information between representational systems, freeing it from its old patterns of discretization and altering the ways in which we can conceptualise it.

Metarepresentation is therefore a great source of cultural variation. People who engage in intense metarepresentational activity are those who are most likely to shift their patterns (Price and Shaw 1998), to provide the stimulus to social reform, to make inventions and discoveries – in short, to introduce the cultural variation that leads to competition and accelerates cultural evolution. Furthermore, this relationship between metarepresentation and cultural variation lends support to the characterization of gifted individuals as highly metarepresentational, because there is also a link between giftedness and nonconformism. A study by Kobe Millet and Siegfried Dewitte (2007b), for example, has demonstrated "a significant positive relationship between the need for uniqueness and general intelligence," suggesting that brighter people will be happier than others to produce novel ideas rather than conforming to the majority opinion.

This behaviour is unusual, because there is evidence for an evolved social learning bias, in preference to the relatively costly process of individual learning, which suggests that the majority of the population will tend to conform to the opinions or practices of either the rest of the majority or a prestigious individual (Boyd and Richerson 1985); "a large body of experimental work in social psychology testifies to the strength and ubiquity of people's reliance on social learning" (Mesoudi 2008: 250). This cohort of less metarepresentational people is just as important for cultural evolution as is the minority of intense metarepresenters. A tendency to metarepresent, to question, to synthesise and review all incoming information, is not the best basis for cultural stability. People who are less able to metarepresent are, conversely, better able to engage in faithful cultural transmission and implementation. It is the majority population's tendency to accept what they are told, to pass it largely unchanged to the next person and to act on it in predictable ways, which accounts for the cumulative and often progressive nature of human culture. Patterns do not always need shifting.

Cultural information relies for its inheritance on human receivers, who vary in a number of ways, including in their tendency to metarepresent. Most of the people, most of the time, will simply accept what they are told, and as a result, they behave as fairly reliable receivers and transmitters of cultural information. Mutations may be introduced, with varying probability, in a variety of forms, depending on the information's content, context and receiver, but on the whole these human receivers are pretty faithful engines of cultural inheritance. Highly metarepresentational thinkers, however, will tend to question and synthesise information before they transmit it to others, and this results in a much greater probability of variations being introduced during the inheritance process.

Metarepresentation, Genetic Fitness and Artefactual Languages

The balance of gifted and modal levels of intelligence within a society is clearly of great significance for cultural evolution, because of the impact that highly metarepresentational receivers have on the inheritance of cultural information. None of us could receive, recombine or transmit cultural information without the metarepresentational pattern-recognition abilities that help us to learn languages and make connections between our concepts, but gifted individuals find all of this much easier than most. Although a later section will argue that education is bound to have an impact on metarepresentational abilities, it is clear that our genes play one of the leading roles in determining variations in human intelligence. We have seen in previous chapters how the advent of artefactual languages brought potential advantages to both cultural information and its receivers, and this section explores in more detail the interactions among giftedness, biological fitness and the shifting cultural context.

It seems reasonable to assume that giftedness, or extreme metarepresentation, might have some significance for human fitness, because of the impact that intelligence has on individuals' responses and behaviours. Indeed, Stanovich and West (2003: 18) have argued that general intelligence is "the single most potent psychological predictor of human behaviour in both laboratory and real-life contexts that has ever been identified." They support a two-process model of human reasoning systems, in which System 1 is a set of autonomous systems in the brain, which operate "in response to their own triggering stimuli and not under the control of a central processing structure," and System 2 "encompasses the processes of analytic intelligence that have traditionally been studied by information processing theorists trying to uncover the computational components underlying intelligence" (2003: 8). Having looked at experiments that separate the two types of rationality, the authors argue that there are situations in which the two are in conflict: where behaving in ways that an individual perceives as being in his own best interests will nonetheless be detrimental to his genes. This distinction becomes obvious when we consider human decisions to commit suicide or, less drastically, to marry somebody whose physical disability precludes her ever having children. They conclude that System 1 evolved to ensure that the organism makes decisions that maximise the survival and/or reproductive probability of its genes, whereas System 2 has evolved as a control system that increases the survival probability of the organism itself, sacrificing the interests of the genes to the interests of the individual where necessary. Consequently, in the few cases where the two conflict, System 1 will instantiate short-leashed genetic goals while System 2 instantiates the longer-leashed goals of the whole organism (2003: 10) – and it is the individual with a high level of analytic (System 2) intelligence who is more likely to prioritise his own interests over those of his genes, tracking what the authors have dubbed instrumental rationality. Lower analytic intelligence, on the other hand, leads individuals to construe tasks in ways that benefit the genes, tracking what the authors call evolutionary rationality (2003: 11).

The spark for the authors' investigations came from the repeated finding in their psychological research that subjects who were more cognitively able did not give the modal response to a problem (2003: 2). They provide several examples of this finding, including the well-known Linda problem, in which each participant is given some information about a woman called Linda and is then asked to rank each of eight further statements about her, according to their probability. The statements include the claims that Linda is active in the feminist movement; that she is a bank teller; and that she is a bank teller and is active in the feminist movement. Because the last statement is a conjunction of the first two, its probability cannot logically be higher than either of the first two, yet 85 percent of participants in the original study decided it was more likely that Linda is both a bank teller and is active in the feminist movement than that she is only a bank teller. From the perspective of deductive logic this is a fallacy, but Stanovich and West cite several investigators who have argued that from the perspective of normal, cooperative, Gricean communication, the participants' reasoning is defensible. When people give us information, we are used to assuming that they intend to be cooperative[1], and

---

[1] Or at least that they intend to play a cooperative game of language: Hurford (2007:270) distinguishes between communicative and material cooperation by analogy with a tennis game, in which competitors "cooperate" only in the sense of playing by shared rules, and not in the sense of helping each other to win.

one of the main ways in which speakers cooperate with one another is by not giving redundant information. In the case of the Linda problem, this means that participants, faced with detailed information about Linda, assume that the experimenter knows a considerable amount about her. They quite reasonably infer, therefore, that the reason the information that Linda is not active in the feminist movement is not included in the statement that she is a bank teller is that the experimenter already knows this to be true (Stanovich and West 2003: 4).

In other words, from the point of view of System 1 reasoning processes, the modal response to the Linda problem is perfectly rational. System 1 comprises mechanisms that support Gricean communication, and participants who make their inferences on the basis of Gricean communication principles should, indeed, produce the modal response. It is only when participants employ the more controlled processes of System 2, which enable individuals to decontextualize and depersonalize the problems that they are asked to solve, that they will be able to focus on the information that is logically present in the statements and ignore the temptation to infer more information than is there. "This system is more adept at representing in terms of rules and underlying principles. It can deal with problems without social content and is not dominated by the goal of attributing intentionality nor by the search for conversational relevance" (Stanovich and West 2003: 9).

What Stanovich and West have discovered is that intellectually gifted, or highly metarepresentational, individuals are more likely to prioritise System 2 over System 1 – and the authors point out that, from a genetic perspective, this might sometimes lower these individuals' fitness levels.

If, however, we take into account not only the biological but also the cultural processes that are at work in these cases, then we can see that the effects of giftedness on genetic fitness are more mixed than these results would suggest. The highly metarepresentational individual is rescued, genetically, by the fact that his tendency to prioritise decontextualized, depersonalised reasoning over System 1 rationality is exactly what he needs in order to use artefactual rather than natural languages. Natural language use, as Stanovich and West point out, is supported by a Gricean prioritisation of communication over interpretation. I have argued that artefactual language use, conversely, demands that we prioritise interpretation and representation, even at the cost of communication. This can only be achieved by an individual who is able to focus on the informational content of the language and the logic of the system in which it is represented, rather than on social factors or any other distractions.

Stanovich and West themselves argue convincingly that this ability is particularly necessary in the modern world. They point out that "knowledge-based, technological societies often put a premium on abstraction and decontextualization, and they sometimes require that the fundamental computational bias of human cognition toward contextualization of problems be overridden by System 2 processes" (2003:14). When we consider our dealings with other people in the workplace, marketplace or financial institutions, we can see that there is often a need to decontextualize the content of our interactions: to provide a retail service even to those customers with whom we should never choose to socialise, for example. Such situations reveal that there is no mesh between the evolved human mind, with its "fundamental computational biases toward comprehensive contextualization of situations" and much of modern life, "with its unnatural requirements for decontextualization" (2003: 20).

Returning to my own work on artefactual languages, however, we can see that there is a very neat mesh between these requirements and the *culturally* evolved nature of modern life. This is no coincidence, for the biologically "unnatural" nature of modern society has culturally coevolved with our methods of dealing with it. Modern society imposes increasing demands for functional rather than social cooperation among its members, but these demands have not sprung up out of nowhere: they have coevolved with the artefactual languages that enable us to meet them. Learned skills in artefactual languages, such as literacy and currency exchange, enable us to cooperate functionally with one another. In a modern context, then, we can see that gifted individuals' ability to prioritise instrumental over immediate genetic rationality might actually, by increasing their personal well-being, have an indirectly beneficial impact on their longer-term chances of survival and reproduction. Metarepresentational abilities enable individuals to acquire and deal more easily in artefactual languages, which enable their users to interact with others who are not members of their own social group. This puts very clever people at a functional advantage in modern societies, because they can use acquired cultural knowledge as a ticket to functional cooperation with a variety of different groups.

In other words, highly metarepresentational individuals might make decisions that would have been genetically irrational in an ancestral context, but they also find it easier than the majority to meet modern society's functional demands. Yet if evolution has no foresight, then how can a capacity that is useful today have survived the rigours of natural selection in the past? The answer is that high intelligence also brings genetically compensatory benefits to individuals, even when they cannot rely on artefactual languages. On the one hand, we have seen that giftedness can bring social disadvantages, because being highly metarepresentational makes it easier for individuals to resist conformist behaviour, and nonconformist behaviour can threaten an individual's acceptance by the rest of the group. In this respect, giftedness would have brought an additional biological disadvantage, because in the ancestral environment it would often have been more difficult for individuals to acquire resources in isolation than in groups. On the other hand, it is also the case that general intelligence strongly predicts future resources (Millet and Dewitte 2007b) – and this means that the brighter someone is, the more she can afford her nonconformist behaviour. She is more able to acquire resources on her own, even if she does not cooperate with others, either because she is not interested or because her nonconformism makes them hostile. If you are clever enough, then you are not so reliant on the low-risk strategy of following the herd – and in any case, individual learning is not such a high-risk strategy for you as it is for someone of lower intelligence.

The effects of high intelligence on genetic fitness appear, then, to be mixed at best. Research indicates, for example, that intelligence is viewed by both sexes as a necessity rather than a luxury when choosing a partner – but only to the extent that it is needed to perform functions like parenting, resource gathering, adapting to change and dealing with competitors (Li et al. 2002). Beyond a certain "sufficiency in intelligence" (2002: 953), any additional intelligence does not endow individuals with any additional sexual gains, and this implies that giftedness is not of any direct sexually selective benefit. As ever in evolution, however, context is all: the advent of artefactual languages provided a newly favourable environment for highly intelligent individuals, because the ease with which they learn and use such languages gives them functional access to different groups of people. The previous chapter has shown how this not only has the potential to increase their resources but also enables them to judge their status against the standards of a different cultural group, altering their self-perception with a consequent benefit for their health.

Nature and Nurture

Cultural information rains down on the landscape of our genetically endowed mental capacities, moulding the paths along which future information must travel, eroding and shaping the patterns of our thoughts and reactions (Price and Shaw 1998). Its impact on our minds is enormous, especially when we are young. But we need to beware underestimating the role of biology in the evolution of culture. It is impossible to imagine what the landscape of a human mind might look like without the nourishment of a cultural downpour, but household pets give abundant evidence of how a nonhuman mind responds to at least some of that input: hardly at all. Cultural evolution depends on both cultural information and the nature of the human landscape on which it falls.

The genetic landscape varies from individual to individual, even though its contours share some universal features. Metarepresentation is not the only innate trait that varies in its expression, despite being universal among humans: we all grow from childhood to adulthood, but there is variation in the heights that we attain; barring physical disability we can all run, but some can run faster or for longer than the rest of us; we all have an eye colour, a blood group and a fingerprint, but again these vary from person to person. This variation results from the unique interaction between environment and genotype that is experienced by each individual organism. You may, for example, have the genetic potential to reach a height of five feet eight inches but as a result of undernourishment only ever grow to be five feet five inches tall. You may be an identical twin, but identical genotype does not mean identical ontogeny or even identical fingerprints: each twin occupies a unique physical space from the moment the fertilised egg divides, and local factors will fill in the details of foetal and subsequent childhood development.

Because nature and nurture interact so complicatedly, their strands must first be untangled before we can trace the roots of any given trait. Most genetically similar children are raised in culturally similar surroundings, so it can be difficult to determine the ultimate cause of their shared characteristics. Even when a characteristic is spread universally among humans of wildly differing cultures, the tangle exists. Twin studies, including observations of identical twins who were separated at birth, can unpick it to a certain extent, but for any given individual, the tangle remains. It is impossible to say with any certainty what *this* individual would have been like if his environment had been different, or what impact his environment would have had on a different individual. This is as true of metarepresentation as of any other varying universal trait, but the next section presents evidence that variations in this ability can be produced culturally, by education, as well as emerging from an individual's genetic potential. We saw in the previous section how the emergence of artefactual languages has changed the adaptive landscape for highly metarepresentational individuals; we shall see in the next section that the acquisition and repeated use of artefactual languages can make individuals more highly metarepresentational.

Metarepresentation and Education

The innate metarepresentation spectrum is part of the reason why some people are less likely than others to move beyond their culture of origin. Put simply, some people have a greater natural tendency than others to question what they are told. Our genetic heritage – the extent to which each of us tends to metarepresent – can help to account for the patterns of cultural evolution. Nature cannot work in isolation from nurture, however, and the level of education that an individual receives will also have an impact on her tendency to metarepresent. For this reason, it is important to emphasise that I do not equate metarepresentational ability simply with intellectual ability as it is measured by intelligence tests. It has been said that IQ tests measure nothing more than people's competence at taking IQ tests, and although this is not fair or accurate, there is a grain of truth in it. A high IQ is partly dependent on an individual's literacy levels, cultural background and experience in taking intelligence tests. Education exposes us not only to new ideas but also to new ways of thinking about them, and it is possible to imagine someone who is far more culturally innovative than others who are innately more metarepresentational than she is, simply because she has been educated to a much higher level than they have. Academic education is often about imparting knowledge of artefactual languages, which open us up to new information and reshape the ways in which we can conceptualise the world.

Research shows, for example, that the ability to overcome natural myside bias is more highly correlated with formal education than it is with cognitive abilities per se. "A large research literature indicates that people have difficulty in decoupling [their thought processes] from prior opinion and belief. So-called myside bias has been amply demonstrated in numerous empirical studies. People evaluate evidence, generate evidence, and test hypotheses in a manner biased towards their own opinions," and they "have difficulty evaluating conclusions that conflict with what they think they know about the world" (Stanovich and West 2007: 226). In two experiments, involving a total of 1,484 university students, Stanovich and West (2007) investigated participants' existing status in four areas: their sex, their smoking and drinking habits and their religious convictions. Participants were then required to evaluate various propositions that were relevant to each of these factors; each proposition was contentious but factually accurate. Although Stanovich and West's find-ings were that higher cognitive ability did not reduce the extent to which participants displayed natural myside bias, Stanovich had found in an earlier experiment with Maggie Toplak (2003) that a longer time spent in university did. Participants' year in university predicted their myside bias scores independently of their ages or cognitive abilities (Toplak and Stanovich 2003: 858). It seems that, because brighter people are better at the decontextualizing that is needed for myside bias, they do bet-ter at avoiding myside bias in experiments where participants are *cued* to decontextualize – but in experiments where no such cues are given, their cognitive abilities are irrelevant and they respond (like all participants) on the basis of System 1 rationality. Educational level might, therefore, be significant because academic education is in effect an apprenticeship in *always* decontextualizing information.

There is a neat overlap, here, between Toplak and Stanovich's work on myside bias, and psychological research into cognitive accessibility. Chapter 6 cited Ying-Yi Hong and colleagues' (2000) research into cul-tural priming, which shows how our responses are influenced by the mental constructs that have been activated by recent use. Hong and col-leagues (2000: 716) argue that "some constructs attain *chronic accessibil-ity*, in part because accessibility is maintained by frequency of use," and I would suggest that this might well be the function of education in cueing decontextualization, thereby decreasing students' levels of myside bias. Similarly, Deanna Kuhn (1991) has suggested that school experience enhances students' critical-thinking skills by giving them opportunities to practise putting their own beliefs to one side in order to think ratio-nally about an issue. It is apparent, then, that academic education is a process of increasing students' representational competence. Not only functional cooperation with other people, but also the effective manipu-lation of modern quantities of cultural information, are enhanced by an individual's acquisition of artefactual languages and all the information that they bring with them.

Cultural Agents and Cultural Diversity

Individuals are prepared by both their genetic heritage and their cultural experiences to act as varying types of receivers for the cultural informa-tion that they subsequently acquire. It is obvious that the metarepresen-tation spectrum is not the only axis along which individual receivers vary, and it is likely that some cultural areas have coadapted to particular types of receivers, including not only more or less metarepresentational brains but also male or female brain types (Baron-Cohen 2003) and so on. Each of us has a particular innate capacity for learning any given portion of cultural skills and facts. Some people are particularly quick to pick up elements of human culture that are underpinned by a relevant system of representation: it is not possible to engage in algebra or statistical analysis, cartography or fluid mechanics, without first having a grasp of the artefactual language in which each is represented, and some people find it very easy to grasp new theoretical information of this sort, to read and write about it and to analyse other people's arguments concerning it. Other elements of human culture are not dependent, in this way, on an artefactual language: it is possible to learn how to build a dry stone wall or to spot-weld, how to make pastry or knit a jumper, with the aid of natural language, imitation and practice, and there are people who struggle with artefactual languages but have natural gifts for learning practical skills of this sort, for honing and applying them with a combination of patience and perfectionism, and for judging the competence with which similar jobs have been performed by others.

Joseph Renzulli's famous three-ring model of giftedness makes this range of abilities particularly clear, and highlights the significance of other factors in addition to intellectual ability or subject-specific aptitude. "Research on creative-productive people has consistently shown that although no single criterion can be used to determine giftedness, persons who have achieved recognition because of their unique accomplishments and creative contributions possess a relatively well-defined set of three interlocking clusters of traits. These clusters consist of above average, though not necessarily superior, ability, task commitment, and creativity," and "it is the *interaction* among the three clusters that research has shown to be the necessary ingredient for creative-productive accomplishment" (Renzulli 2005: 259). In Renzulli's Venn diagram, where the three overlapping circles represent ability, commitment and creativity, gifted behaviours occur in the central area where all three circles meet.

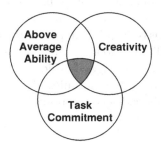

It is important to emphasise that Renzulli's theory applies to the gifted behaviours that individuals might evince in particular areas of their lives, rather than to the individuals themselves. People are not equally motivated in every area of their lives, even if they are obsessively well motivated in certain areas. People whom we describe as creative individuals may have areas of their lives in which their behaviour is quite mundane. Above-average ability in one area offers no guarantee that the same person will have even average ability in another. "All three rings, working together, are important for creative-productive giftedness, and all three rings exist on a continuum of both accomplishment and potential" (Professor Joseph Renzulli, personal communication). No combination of two without the other one will suffice. "If a child has remarkably high motivation and creativity in an area such as robotics, for example, but does not posses high ability in this area (forget IQ), then there is a limit on the quality of product that she or he can produce. The high motivation and creativity might overcome some limitations in technical ability, and even stretch the child to become more technically proficient than we might expect, but ability always puts a limit on accomplishment regardless of the strength of the other two sets of characteristics" (Professor Joseph Renzulli, personal communication).

Not only innate ability, then, but also creativity and motivation are part of the genetic landscape onto which culture falls. Cultural evolution is primed by this genetic diversity. Across different cultural areas there is a continuum of potential in these three factors, not only within each individual but also across different individuals. What this highlights is the significance for cultural evolution of the range of individuals within any group. Culture is the product of interactions among many individuals, each with a unique profile of abilities, motivation and creativity. Renzulli (personal communication) asks us to imagine what might happen if we took our highly motivated child, with his "remarkably creative idea for a new kind of robot," and teamed him up with someone who is less creative but has the technological skills to pursue the creative idea. "Many of the great accomplishments in all areas of human productivity are the result of people who bring different competencies to the table." The genetic variation among humans is a significant part of the landscape onto which cultural information falls, and so is the information that has fallen there in the past, along with the languages in which it was represented. Culture is the product of interactions between cultural information and its varying human receivers, and their diversity is bound to affect the course of its evolution.

PART V

THE EXPRESSION OF
CULTURAL INFORMATION

12

Aspects of the Cultural Ecology

The origins of culture lie in the biological evolution of organisms with the capacity to receive the information on which culture is based. Our ancestors were biologically preadapted for the acquisition of natural language, which evolved culturally as a method of enhanced communication between members of a cooperating species. The acquisition of natural language turns each member of our species into a receiver for the cultural information that is represented in her native tongue. Information can be inherited only by receivers that can discretely represent its variations, and natural language delivers the ability to discretize information in the same way as the speakers from whom we have learned it. Although natural language has structural features which indicate that it evolved primarily as a method of communication, and its practice reveals a range of representational deficiencies, its representational capacity was sufficient to enable our ancestors to share an increasing amount of cultural information. Eventually, when this information exceeded their collective cognitive abilities, they found themselves prepared, by the same biological preadaptations that had made them ready for natural language, to acquire artefactual language. Members of this metarepresentational, cooperative species were able to learn ever-more-efficient methods of representation. The use of these novel representational systems was not restricted to kinship groups, but rather facilitated the formation of functional links between individuals who were socially unrelated to each other. Variations among members of this species, especially in their tendency to metarepresent, have been key factors in the course that cultural evolution has taken since the advent of these natural and artefactual languages.

This chapter explores the artefactual and behavioural results of cultural information's expression by its human receivers. It examines the factors that affect how human receivers react to the cultural information that they acquire, discovers that the "phenotypic" effects of cultural information sometimes form recognisable phenotypic units, and shows how the function and reproduction of each unit will usually be the expressions of different portions of cultural information. It considers the impact that aspects of the cultural ecology will have on the directions taken by cultural evolution. Finally, it highlights the ways in which our everyday language classifies cultural units into hierarchical clusters. This suggests that there is, at the very least, a pattern of cultural diversity that provides an explanandum for the project of cultural taxonomy on which Chapter 13 will embark.

Cultural Units

In order to understand what cultural agents do with the information that they receive, it will be helpful to start by looking at how biological information is expressed.

An organism's genes are expressed by the organism's cellular machinery, in ways that influence the survival and dissemination chances of both the genes and the cellular machinery. These include the physical development and maintenance of the organism, effects on its behaviour and the creation of its descendants. The end result of these effects is the preservation and dissemination of that genetic information and its inheritance mechanisms.

A virus's genetic material is expressed by its host's cellular machinery, in ways that influence the virus's chances of survival and dissemination and will often also have an impact on the host's chances of survival and dissemination. These include both physical and behavioural effects on the host, as well as the creation of the virus's descendants. The end result of these effects is the preservation and dissemination of the virus's genetic material, which will often be at the expense of the host's health.

Cultural information is expressed by human language users in a variety of ways, which influence the cultural information's chances of survival and dissemination and may also have an impact on the human receiver's survival and reproductive hopes. These include the immediate transmission of the information to another receiver (with or without mutations); a creative process of metarepresentation, in which links are made between the new information and other information that the receiver has acquired; forgetting; and putting the information into action, which may produce either behavioural or artefactual results, or indeed both. The end result of these effects is the preservation and dissemination of the cultural information, which will often have some impact on the human receiver.

From this brief sketch, it is clear that cultural information is more like the genetic material in a virus than it is like an organism's genes. Viral and cultural information rely for their detection, expression and dissemination on receivers that are physically separate from themselves, and they have only incidental effects on their receivers' reproductive chances. For this reason, there is no real cultural equivalent of an organism. Yet even a virus produces delineable phenotypic effects, via the cellular mechanisms of its host. It is the aim of viral taxonomy to identify and classify these effects: the effects that the virus has on its host, the character of the virus during the virion stage of its "life" cycle, and so on.[1] Virologists might sometimes struggle to assign a virus to a species or to understand its relationship to other viruses, but they know a virus when they see one and they can delineate its effects, whether these take the form of a literal, physical unit or the form of physical or behavioural effects on its host.

Cultural information, similarly, has phenotypic effects on its environment. As in biology, what we encounter most directly in the cultural world is very often not information but the effects of information: the physical and behavioural results of its receivers' reactions to it. Walking down the high street, we see not fashion designers' drawings and instructions but the different items of clothing that people are wearing. We hear music, even though we have no access to the sheet music or electronic programmes on which it is based. We buy a pasty for which we do not know the recipe. We catch a bus about whose design we know nothing. The distinction between information and its effects is as crucial for culture as it is for nature. In both evolutionary arenas, information is selected not directly but by means of its effects: the information builds and operates phenotypic features whose ultimate function is to maintain and propagate that information in a particular environment (Cloak 1975a: 170).

The most cursory glance around human culture reveals a selection of delineable units, in the same way that nature confronts us with individual organisms and virions. We can see skirts and cars, books and cups, CDs and suitcases, cakes and trumpets, pictures and gates – innumerable individual artefacts whose boundaries are clear for all to see. We can also observe less tangible, behavioural effects that nonetheless are fairly clearly bounded: the pattern of a school assembly; the etiquette of joining

---

[1] A virion is the viral particle as it exists in its capsid outside the cell.

a queue; the habit of eating with the mouth closed; the disapproval of adultery; the size of one's personal space.

Cultural units of this kind should not be compared too closely with biological organisms. Not only is cultural information viral, in the sense that it is separable from the mechanisms of its interpretation and expression, but portions of a cultural unit can also be reproduced independently of the whole, in a way that a portion of a biological organism cannot. We might draw some sort of comparison between, say, an amputated limb and an incomplete phrase that would never stand on its own as a piece of music, but the snatch of music could still be reproduced in isolation from its original unit, regardless of its incompleteness, in a way that the limb could not. Still, there is a sense – useful despite its limitations – in which we can say that culture, at the phenotypic level, falls naturally into distinct units that we can identify and investigate.

Cultural Receivers

Cultural information's expression in these units depends on the existence of appropriately skilled and knowledgeable human receivers. Renzulli (2005) has drawn our attention to the fact that culture is the product of many individuals working together in many different areas. It is facilitated by their capacities to form social connections with one another using natural language, and to make functional connections with one another using artefactual languages. Although discussions of cultural evolution often focus, quite naturally, on human cognitive capacities, it is fruitful to remember that the agents of cultural evolution are not just human brains but human individuals taken as a whole. Each person has a particular genetic endowment, which includes both species-general and individual-specific physiological and psychological traits. Variations will include the degree to which he naturally conforms or metarepresents, systemises or empathises, has good hand-eye coordination or particular physical or learning disabilities, and so on. In addition, each person will have acquired a particular set of languages and skills, which will affect not only how he responds to new information but also how (and indeed whether) he is capable of implementing the cultural information that he encounters. His existing knowledge and skills are partly the result of individual learning and biology and partly the result of earlier cultural learning.

This range of skills, knowledge and opinions is crucial for the interpretation and expression of any new cultural information that an individual encounters. Physical skills are as vital for this process as intellectual knowl-edge, for information cannot be expressed in isolation from appropri-ately skilled human agents: in addition to knowledge *about* something, the ability to *do* something is also needed. Although I know in theory how a flute should be played, I remain unable to produce a note from one, and an explanation of how to cure Alzheimer's disease would be useless if it described actions of which we were incapable, just as much as if it were written in a language that we could not decipher. If knowledge of the relevant language is needed for the transmission of information, then possession of the relevant skill is needed for its expression.

Human skill, in other words, enables cultural information to have its effects: to give rise to artefacts or behaviour. But skills are at least partly dependent on individual rather than social learning, in that they are not transferable between people.[2] You can share information about best practice, but you cannot share the ability to put it into practice. Although it can be argued that there is a continuum between the processes of social and individual learning – that the two are mutually dependent and they interact reciprocally over time to strengthen one another (Salomon and Perkins 1998) – there remains a fundamental distinction between these two interacting processes. On the one hand, there is an individual learner, not only acquiring new information but also developing new skills; and on the other hand, there are processes by which that learner can acquire the transferable aspects of his learning from external, social sources. As any frustrated teacher knows, not all of the things that a pupil needs to learn are actually transferable. You can give all the hints and tips in the world to somebody who is unable or unwilling to put them into practice, and still he will not acquire the skills of which you are capable. An athletics coach can pass on his knowledge to his team, but he cannot transfer his musculature to them. A clarinet teacher can pass on her musical insights to a pupil, but she cannot pass on her ability to put them into practice. Although how-to knowledge is a transmissible bit of information, physical ability is not transmissible but can develop only through individual practice. This is why, no matter how many books they may read, seminars they may attend or videos they may watch about a particular subject, some people will never be as skilled in that area as others. Their individual learning potential (innate ability, motivation

---

[2] Thanks to If Price, with whom the thought originated that reading music, for example, is ecological from a cultural point of view.

and creativity in this particular area) is more limited, and therefore the end results are less impressive.

One complicating factor, here, is that there are two separate sets of information associated with any given artefact: the information about its use and the information about its reproduction. In the case of behaviours, similarly, it is not always the case that someone who can do something himself can also teach somebody else to do it. It is perfectly possible – indeed, it will often be the case – for someone to have one portion of information and one set of skills but not the other: to be able to use but not to copy an artefact, for example, or conversely to be able to reproduce but not to use it. I have a limited ability to play the mandolin and my husband is able to make one, but I cannot make a mandolin and he cannot play one.

My mandolin's dual cultural role is not very different, in fact, from the dual role that an organism or virion plays within nature. An organism or virion is an entity that is built on the instructions of a particular portion of genetic material, using the available chemical tools and materials. In this role, it is the physical expression of a particular portion of information. But a genome contains instructions for the organism's or virion's behaviour as well as for its construction – and this behaviour is the expression of a different portion of the genetic instructions. In relation to this portion of its genetic instructions, the organism or virion itself is not their expression but is that which provides the necessary materials and actions for their implementation.

Similarly, my mandolin is built according to a particular set of cultural instructions, using the available physical tools and materials. In this role, the mandolin is the physical expression of a particular portion of technological information, produced via a combination of the necessary human skills and the requisite physical materials. In addition, the mandolin itself provides the material through which a different portion of information – musical, rather than woodworking information – can express its instructions. In relation to these musical instructions, the mandolin is not their expression but the material that enables them to be put into action. Mandolin music might be seen as containing instructions for the mandolin's "behaviour," and the phenotypic effect of those instructions is not a mandolin but mandolin music.

Longevity is not identical with fecundity. What I provide for my mandolin is a mechanism that enables it to "live," or more accurately, to function. What my husband provides is a replicative mechanism: the means of its reproduction. Left in the attic, my mandolin would be no different from a vacated spider's web. Left alone with me, it would function but never be reproduced, like a sterile organism that lives a long and healthy life. Left alone with my husband, it might well be reproduced but it would never function. There is no comparison for this in the biological world, because if a virus or organism does not "function" then it is dead, and dead things cannot reproduce. An organism or virus's genetic material controls its operation and reproduction, but the two facets can never be separated in the way that they are in culture, because genetic information comes as a package, whereas we can pick up different portions of cultural information, such as those for operating an artefact and those for making it, at different times in our lives.

Environmental Factors

As well as the physical and cognitive skills of locally available humans, other aspects of the cultural ecology are external to humans, such as the local physical environment. This includes both natural factors, like climate and geography, and cultural factors like the impact of physical workspaces on human interactions and hence on cultural evolution. Price (2007, 2008), for example, draws attention to the reciprocity of the ways in which the shape of physical work spaces both reinforce and are determined by existing beliefs and practices. Mark Pagel and Ruth Mace (2004) have highlighted the impact of the wider physical environment on cultural diversity. Like biological species distribution, they have found that the pattern of cultural distribution – as indicated by the use of different natural languages – is denser nearer the equator and sparser further north. Pagel and Mace suggest that lower levels of cultural diversity in northern latitudes may be a product of the physical environment, which is relatively unproductive and hence drives individuals to cover larger geographical areas in the struggle to survive. Cultures and languages tend to be homogenized by this movement of people (2004: 276), whereas Pagel and Mace (2004: 278) suggest that when resources are plentiful, cultural diversity is maintained by a drive to withdraw from larger groups in order to control defensible resources, and by social and behavioural practices that support cultural identity, coherence and cooperation within groups and conversely cause barriers to the flow of genes between groups, so that cultural transmission becomes predominantly vertical with respect to genetic generations.

In these ways, the physical environment influences the differential transmission of cultural information and the ways in which it is interpreted and expressed, but it is important to bear in mind that the prime movers in cultural evolution are human agents, living and working together as a result of the natural and artefactual languages that they have acquired.

Technological Evolution

Once cultural information has been expressed in phenotypic effects such as artefacts or behaviours, several factors affect the success that those cultural units might enjoy in the competition for human attention. The history of technological innovations provides several fascinating illustrations of this process, and in this section I briefly describe the development of two different technologies, before drawing out the evolutionary features that they have in common.

Example 1: A Brief History of the Mobile Phone
The technology that underlies mobile (cell) phone use was proposed as long ago as 1947, by engineers at Bell Labs. The suggestion was that service areas should be divided into hexagonal cells, and that at the corners of the hexagons there should be towers with directional antennae that would transmit and receive into three adjacent cells. The proposed technology did not exist at the time, and it was not until 1983 that first-generation mobile phones came onto the market. First-generation phones used analogue transmission between base stations in a network of cells, and there were protocols for an automated handover between cells as the phones moved around. Although genuinely innovative, this analogue transmission was relatively unreliable and produced a lot of static and noise, and first-generation phones were originally restricted to permanent installation in vehicles. Even when they became genuinely mobile, these phones were the size of a briefcase and prohibitively expensive. For much of the 1980s, they remained a status symbol rather than a truly convenient tool.

As technology improved, a second generation of mobile phones emerged. These phones used quieter, faster digital technology and were much smaller than their forebears. This was partly the result of technological developments like smaller batteries and more energy-efficient technology, but it was also helped by increased usage of the networks: instead of widespread base towers demanding higher transmission power, the network of towers became denser, increasing both coverage and reliability. In addition to the smaller, more attractive handsets, lower prices helped to contribute to the popularity of second-generation phones.

Third-generation technology allows images and videos to be transmitted, and high-speed Internet access as well as texts and voice data. In the future, it is likely that information transfer, and in particular Internet capacity, will become even faster and more affordable.

Example 2: VHS and Betamax

Sony introduced their Betamax recording technology in 1975, followed in 1976 by JVC's rival VHS technology. The competition between the two is legendary. Was VHS successful in the end, despite inferior quality, because it was adopted by the U.S. porn industry? Many people believe this version of events, and the story of how the porn industry's influence triumphed over technological quality has attained iconic status. According to journalist Jack Schofield, however, the real story is both more complex and more interesting than this.

Schofield (2003) highlights the fact that consumers in the 1970s and 1980s were not choosing only between a VHS and a Betamax videotape, or even only between a VHS and a Betamax videocassette recorder. Schofield introduces a marketing concept, the whole product, to explain what he means. He points out that the technological aspects of a product are a small, and often uninteresting, part of a consumer's decision between two products. In addition to the core product, she must also buy into the product's infrastructure. This includes the product's availability and price, the availability and price of the associated infrastructure, and so on. "In marketing terms, 'the core product' – such as a car, a computer, or a video recorder – is just the start. You have to add on all the things like reliability, service and support (the expected product), its expansion capabilities (the augmented product), and its potential for future development (the potential product) to get 'the whole product.'" People take into account the whole product when they make buying decisions in the real world, and consequently it is highly misleading to make simple comparisons between the technological features of core products.

The concept of the whole product, according to Schofield, gives us a more convincing explanation of why VHS triumphed over Betamax. Not only did VHS offer a cheaper and wider choice of hardware, and a cheaper and more easily available range of tapes, but Betamax's onetime dominance of the market was undermined when Sony "got one simple decision wrong. It chose to make smaller, neater tapes that lasted for an hour, whereas the VHS manufacturers used basically the same technology with a bulkier tape that lasted two hours. Instead of poring over the sound and picture quality, reviewers could simply have taken the systems home. Their spouses/children/grandparents and everybody else would quickly have told them the truth. 'We're going out tonight and I want to record a movie. That Betamax tape is useless: it isn't long enough. Get rid of it'" (Schofield 2003).

It turned out that consumers would rather sacrifice picture quality if it meant increasing the available recording times. Betamax, says Schofield, could not offer consumers what they wanted: the ability to record a whole film in their absence. Betamax did extend their playing times, but never enough to catch up with VHS. The reason why VHS won the marketing battle, according to Schofield, is that the whole product that Betamax was offering did not do what people wanted it to do, at a price that they were willing to pay.

An interesting footnote to this story is that Betamax might have lost the battle for the domestic video-recording market, but on some accounts, Sony nonetheless won the financial war. Having withdrawn Betamax machines from the consumer market, "Sony retargeted Beta videotape recording to a market that valued video quality more than tape length. The result: Sony dominated the professional and broadcast market from the 1980s well into the 1990s with its Betacam, Betacam SP, and Betacam Digital series of products" (Howe 2006). According to the industry analyst Carl Howe (2006), Sony made ten times more annual profit from its professional video technologies for the broadcast industry than any manufacturer of VHS videocassette recorders did, even at their peak.

Technological Evolution as the Expression of Evolving Cultural Information
In both of these examples, we can see some typical features of technological evolution, all of which are a population-level product of the interaction between cultural information and the cultural ecology. First, there is the coevolution of technology with infrastructure: just as trains rely on the rail network and cars on a network of metalled roads, so mobile phone use is dependent on the associated cellular network, and sales of video-recording technology succeeded the widespread adoption of television sets. This infrastructure is a vital element of the whole product, and if it is not in place then the core product cannot hope to succeed.

There will be times, for example, when an innovative technology has to compete with an existing alternative whose infrastructure is so well established that there is little incentive to invest in what would be needed to support the new technology. In other words, the new technology might be better in all respects than the old, but the new technology will not be widely adopted until the old infrastructure needs replacing: if a success- ful existing technology, seen as a whole product, is so supported by the cultural niche that it occupies that there is no room for an alternative, then a kind of cultural inertia can keep an inferior technology dominant in a cultural niche, long after a different sector has produced better solutions to a particular need. There was a lag, for instance, "in manu- facturing plant design as electrical power displaced steam in the early part of the Twentieth Century. Steam boilers were large plants and, like water mills of an even earlier economy, drove rotating overhead shafts. The constraints of the transmission system dictated the geography of the production process. Electrical transmission and electric motors allowed power to be localized, but there was a twenty year lag before manufactur- ing plants exploiting the new capability became commonplace" (Price 2007: 6). Something similar happened when office geographies that were inherited from an earlier technological era failed to adapt adequately to the arrival of distributed computing and wireless technology (Aronoff and Kaplan 1995).

Cultural innovations, then, are partly dependent for their success on the cultural context. The existing cultural information has found an eco- logical niche to exploit, and if novelties are to succeed then they may have to find a different niche. One of the ways in which this can happen is when early technological evolution takes place in an environment that is somehow isolated from competing technologies. Because novel tech- nologies will find it difficult to carve out niches in cultural "ecosystems" that are broadly stable, they may have to evolve, instead, in smaller iso- lated populations before they can challenge the mainstream (Rothschild 1992). Often, this will mean that they come into existence as playthings of the rich: expensive prototypes or demonstrators, made in small num- bers, and in any case impractical for the general population, because the appropriate physical and marketing infrastructure is not yet in place.

A very early example of this process can be seen in the evolution of the spoked wheel. Although a properly made spoked wheel is adaptively superior to a solid wheel, the complexity of its engineering means that if any part, or relation between the parts, of a spoked wheel is not correctly executed, then the wheel will not be superior but definitely inferior to a good solid wheel. Ted Cloak (1968: 7) cites an informant who told him that while "a properly made wheel will last five years under heavy loads in the tropics; an improperly made wheel will last about two months." Critically for the eventual evolution of spoked wheels, however, Cloak

(1968: 8) explains how the earliest examples did not have to last five years, because they were used exclusively on chariots: "They had merely to carry at the most two men, and last through one year's fighting-season or perhaps only one battle."

Another common feature of technological evolution is that once a small, isolated population has evolved, its success will be driven more by market forces than by any absolute technological comparisons. Fitness, as ever, is a relative concept. No matter how superior a novel technology might be, and what its advantages are to the human population, its widespread adoption is unlikely to happen until its price has dropped, or its convenience increased because a whole-product infrastructure has been developed. Marketing departments do their best to manipulate how the public perceives each product, using strategies like advertising and celebrity endorsements. Ultimately, however, its success will be determined by its adaptation to its ecological niche. What is the public looking for in this sort of product? The answer is not always easy to predict. In the case of mobile telephony, size matters but so does network coverage and ease of use. When it comes to video technology, it turns out that domestic customers value quantity over quality, whereas professionals feel just the opposite. This was why VHS technology was better suited to the domestic market and Betamax to industry: each was adapted to a particular cultural niche. As Schofield (2003) emphasises, the key to success lies not in the absolute technological specification of a core product but in the fit of a whole product to the local human ecology: artefact-level technological evolution interacting with population-level adaptive pressures.

Cultural Clusters

One result of these population-level pressures is that cultural units tend to form recognisable clusters. Among the individual differences and similarities across a range of artefacts and behaviours, we can recognize particular groups that have morphological or behavioural similarities in comparison to other such groups. We can see, for example, that some aspects of culture are behavioural and others artefactual. We have higher-level names for types of artefacts, such as *tools* and *books, building materials* and *vehicles*. Within these categories, we can further break the items down, so that, for example, there are carpentry tools and cooking utensils, mechanical tools and writing implements. Looking more closely at the writing implements, we can see that some are typewriting tools and others are handwriting implements. There are different kinds of handwriting implements, which make their marks in various different materials such as chalk and ink, crayon and graphite. There are different kinds of pencils, including some that encase a rod of graphite in wood (or more recently in recycled materials like paper or plastic) and others that use a plastic or metal mechanical device to hold it in place. There are different grades of graphite, suitable for using on different kinds of surface, and for this reason different types of pencil will be used by carpenters and artists, writers and seamstresses.

Similar distinctions can be made among types of behaviour: we call some behaviour leisure and other types of behaviour work, and there is a distinction, for example, between manufacturing and service work. Within service industries, there are differences between financial, legal, retail, catering and caring professions. There are many different caring professions, such as social work and child care, counselling and oncology, teaching and nursing. Nursing care involves different sorts of behaviour, including infection control and emotional support, administration and medication, personal care interventions and checking vital signs. Vital signs such as temperature, blood pressure and pulse will be checked using a variety of techniques and equipment.

At an anecdotal level, then, we can observe groups of cultural activities and artefacts that have obvious similarities and form some sort of cultural cluster. Our very vocabulary performs a taxonomic function at this anecdotal level. Some cultural units can be grouped, at least at the folk-scientific level, into clusters with clear-cut boundaries. Both artefacts and behaviours exist that are stable enough to be characterised by a few common defining properties: there would be little problem in answering the question, for example, whether a particular artefact were or were not an umbrella; or whether a particular behaviour were or were not an act of writing. Other cultural clusters are not so clear cut: like viruses, they form polythetic classes, characterized by a combination of properties rather than by a single common defining property.[3] A polythetic approach is particularly relevant for entities "that undergo continual evolutionary changes and show considerable variability," resulting in clusters with "hazy boundaries" (Van Regenmortel and Mahy 2004).

Polythetic Classes and Family Resemblance
Many commentators (most famously, Needham 1975) have drawn a comparison between polythetic categories and Ludwig Wittgenstein's (1953)

---

[3] There is a more detailed discussion of polythetic viral classes in Chapter 13.

characterization of game-playing activities, among other cultural phenomena, as sharing a family-like resemblance (*Familienähnlichkeit*) rather
than having any one thing in common. When we consider the nature
of a game, says Wittgenstein (1953: sec. 66), we might wish to produce
a clear-cut definition, but if we "don't think, but look!" then we shall
see that there is actually no one characteristic that all games have in
common. Only some games are competitive; we should not describe all
games as amusing; some games are based on chance and others on skill;
and so on. Thus we can see that games form a family, or a twisted thread
of fibres: "the strength of the thread does not reside in the fact that some
one fibre runs through its whole length, but in the overlapping of many
fibres" (Wittgenstein 1953: sec. 67).

Although polythetic classes are obviously reminiscent of Wittgensteinian families, there are differences between the two. "Wittgenstein
is primarily concerned with the terms of ordinary, prescientific language
in which we do not worry too much about precise definitions and delim-
itations" (Schwartz and Wiggins 1987). When he describes the family
resemblance between games, he is referring to how we use words and
concepts in everyday conversation (Chaney 1978: 139) and drawing
attention to the fact that our concepts remain usable even when their
edges are blurred (Wittgenstein 1953: secs. 69–71). Although it is possible to draw a sharp boundary around a Wittgensteinian family of items,
defining it as the disjunction of a set of properties, or the logical sum
of various concepts, it is also possible to define it in such a way that its
extension does not have a fixed frontier. In fact this, argues Wittgenstein
(1953: sec. 68), is just how we do use words like *game* and *number* in
everyday speech.

Polythetic concepts, in contrast, refer not to our everyday language use
but to the objects that we are seeking to classify. They are scientific terms,
as precisely defined and delimited as possible (Schwartz and Wiggins
1987). In polythetic classes, each member has many of the category's
defining attributes, and each attribute is possessed by many members,
but there is no attribute that is possessed by every member (Livesley
1985: 354).

On the one hand, it is perhaps unfair to expect Wittgenstein to provide
much of an insight into scientific definitions, when he was talking only
about the everyday ways in which we use language (Schwartz and Wiggins
1987). On the other hand, the comparison between Wittgensteinian
families and polythetic classes does highlight the importance, for cultural
evolution, of distinguishing between quotidian and scientific levels of

description. At an anecdotal level, we can agree with Wittgenstein that it is possible to talk meaningfully about cultural clusters, even when we are unable to verbalise a precise definition of them. But this does not entail that there is no real, underlying pattern of taxonomic relationships between cultural clusters and their constituent units. As Richard Chaney (1978) has pointed out, Wittgenstein's intuition that a concept might be blurred and inexpressible, yet remain usable, is an intuition about everyday language use. In contrast, the claim that many cultural clusters will – like viral species – form polythetic classes is a claim about all of the concrete objects and observable behavioural patterns that make up human culture. Despite the fact that our everyday talk about these cultural units is not always undergirded by precise definitions, a cultural taxonomy of the units themselves remains a possibility.

Human culture is the product of interactions between human agents and cultural information. Cultural information is expressed in populations of cultural units, which are roughly classified, by our everyday cultural understanding, into hierarchically arranged clusters of similar types. Population-level outcomes result from changes in cultural information, from the constantly changing ecology of human agents, and from the existing cultural context. The resultant cultural clusters invite analysis, providing the explanandum for cultural taxonomy.

13

Patterns of Cultural Taxonomy

In addition to the ecological pressures that the previous chapter described, this chapter will argue that patterns of cultural taxonomy can be understood as the product of interactions between cultural information and its inheritance mechanisms. This mirrors the way in which patterns of biological taxonomy can be understood as the result of ecological pressures as well as of the interactions between genetic information and its inheritance mechanisms. The mechanisms that are responsible for the reproduction of genetic information have a significant impact on the spread and composition of biological populations: in nature, as the next section shows, the very concept of a species is affected by the ways in which different organisms reproduce. Because the mechanisms of cultural reproduction also vary, it seems reasonable to expect that they, too, will have taxonomic implications. This chapter explores the impact of cultural inheritance mechanisms on patterns of cultural diversity.

Biological Taxonomy

In sexual reproduction, genetic material from each parent is recombined and expressed in a new phenotype. The variation that is introduced by recombination is advantageous, but if too much variation were permitted then too many unviable individuals might be produced. For this reason, species boundaries limit the amount of genetic variation in each population (Bock 2004a: 180). Groups of interbreeding individuals are genetically isolated from other such groups by a variety of species barriers, including both physical and behavioural characteristics and gene-level factors. The result of these species barriers is that the taxonomy of sexually reproducing species is hierarchical, or perfectly nested: every pair of taxa is either wholly separate from each other, or one is contained wholly within the other; there is never any partial overlapping. The evolutionary relationships between species can be represented on phylogenies, which are the family trees of taxonomy: speciation (the splitting of a lineage) is shown as a branching on these trees, just as in a more familiar context the birth of a new child is shown as a branching on a person's family tree.

Organism versus Gene Taxonomy

The transmission of information from one generation to the next always depends on a receiver that can link the incoming information to the intended output, expressing the meaning of the information in an appropriate way. When species barriers prevent the intermingling of information, each receiver can make sense only of the information in its own pool, and branching events are always final. When information is always passed on in a package, the evolutionary history of the package will be the same as the evolutionary history of any portion of the information it contains.

In situations where the species barriers are more porous, however, information can sometimes seep from one pool to another, creating a new package of information, all of whose contents no longer share the same evolutionary history. When this happens, the map of whole packages' history and relationships will not accurately reflect the history and relationships of each portion of information within the packages. For this reason, although much evolutionary history can be captured by hierarchical, tree-like classifications, "in other taxa or at other levels, reticulation may be the relevant historical process, and nets or webs the appropriate way to represent what is a real but more complex fact of nature" (Doolittle and Bapteste 2007: 2048).

There is an interesting parallel, here, between the situation at the level of information and what we find when we look at its effects. Phenotypic effects can be extended (Dawkins 1982) beyond the boundaries of organisms to locations in the environment (e.g., a spider's web) or even in other organisms (e.g., a parasite's effects). When the phenotype is extended in this way, so that some of an organism's phenotypic characteristics stem from its own genes and others result from parasitic infection, we can see that there is more than one strand to the causal history of its phenotype.

Similarly, when species barriers are porous, and genetic information is not necessarily transmitted in fixed packages, we can see that there is more than one strand to the evolutionary history of a genotype. Phylogenetic trees have their roots in the assumption that organisms' history and relationships can be mapped by tracing the evolutionary history of their genotypes – but if there is more than one story to be told about the contents of any given genotype, then this assumption is invalid. At least in some taxa, and at some levels, the evolutionary history of a species of organism will not match the evolutionary history of every element of that species' gene pool. Given sufficient data about the history of genes, we could, in theory, "ask which genes have traveled together for how long in which genomes, without an obligation to marshal these data in the defense of one or another grander phylogenetic scheme for organisms" (Doolittle 1999: 2128).

Reticulate Prokaryotic Evolution

This is especially true for prokaryotes, the bacteria and archaea whose genetic material is not contained in a cell nucleus (unlike that of the rest of the organic world, the eukaryotes). Doolittle and Bapteste (2007: 2046) argue that inclusive hierarchies do not emerge naturally, with any consistency, from prokaryotic data. Rather, the horizontal transfer of genes across prokaryotic species, by mechanisms like transduction, conjugation and transformation, act to convert potential trees into reticulate networks.

This is significant because, although we tend inevitably towards a rather anthropocentric view of the world, the reality is that prokaryotes' evolutionary history is 2 billion years longer than that of the eukaryotes, and they are of course still enormously abundant today: it is estimated that the number of prokaryotes on earth is in the region of $4 - 6 \times 10^{30}$ cells (Whitman, Coleman and Wiebe 1998). The earth is laden with the diverse fruits of their long evolutionary history: everywhere from 130°C hydrothermal vents (Nee 2004) to $-10°C$ ice cream (Weinzirl and Gerdeman 1929), as well as across an astonishing spectrum of pH levels, salinity levels and atmospheric pressures, prokaryotes survive via a range of different metabolic mechanisms (see Railsback 2007 for a review with citations). Their evolution has taken them in myriad different directions. Although the number of their species is unknown, it is clear that the microbial world is vast, significant and unfathomably complex: an estimate of 10^{10} extant taxa is on the one hand "a barking mad upper estimate" (Professor Tom Curtis, personal communication), which on the other hand implies a very slow rate of net evolution, given how long the prokaryotes have had to evolve (Curtis et al. 2006).

In other words, although it can be tempting for us to assume that organisms have moved on from their primitive prokaryotic origins, the reality is that two-thirds of evolutionary history took place before the eukaryotes appeared and that prokaryotes survive today in a number and range that should humble the proudest eukaryote. This means that the pattern of prokaryotic evolution cannot be dismissed, either as something that was relevant only in the earliest days of life on earth, or as trivially infrequent reticulation events. Rather, it tells us that hierarchically branching trees might not always be the best means of mapping evolution. Although the phylogenetic tree provides "an appropriate model for many taxa at many levels of analysis" (Doolittle and Bapteste 2007: 2048), it is not applicable to the reticulate evolution of information that can escape from its original packaging.

This finding is of obvious relevance to cultural evolution, in which "reticulation is rampant" (Gray, Greenhill and Ross 2008: 10). Only once we acknowledge that reticulation also occurs across the natural world can we begin to make useful comparisons between nature and culture rather than be misled by our assumptions. The significance of biological reticulation is that it requires us to ask separate questions about the phylogenies of individual genes and the phylogenies of organisms, and we shall see that a comparable distinction will often need to be made in culture.

Viral Taxonomy

The distinction between gene and genotype histories is also relevant to the taxonomy of viruses, and given the aspects that viral and cultural information have in common, it will be interesting to finish our tour of biological taxonomy with some insights into the viral situation. Viruses are intracellular parasites: they do not have their own cells but wrap their genetic material in a coat of protein known as a capsid. As well as protecting the virus's genetic material, the capsid provides a key to the host's cells, on which the virus depends for its growth and reproduction. Viruses are at an advantage in this respect, because their short generation times combine with relatively high mutation rates to facilitate swift adaptation to changes in the host environment. Indeed, viruses must be able to evolve more quickly than their host cells; otherwise, the host cells would be able to adapt or evolve to avoid infection. In the face of host immunity, viruses need either a steady supply of new hosts (in the form of children, for example) or the ability to cross over into a new kind of host.

Viruses' dependence on their hosts means that there is a certain degree of coevolution between the two, but obviously the relationship is not straightforward enough for us to be able to map organisms' taxonomy onto the taxonomy of the corresponding viruses. In contrast to the higher viral taxa, which are universal classes whose members share a set of necessary and sufficient properties (Van Regenmortel and Mahy 2004), a virus species is defined as "a polythetic class of viruses that constitutes a replicating lineage and occupies a particular ecological niche" (Van Regenmortel 1990).

It is worth unpacking this definition. Members of a polythetic class will share several properties with other members, but will not necessarily all share a single, defining property (Ball 2005: 3). Virus species are therefore characterized by a combination of properties rather than by one single, defining property. This approach has the advantage of being able to accommodate individual viruses that lack one or more of the properties that are normally considered typical of a viral species – an advantage that is particularly significant when classifying entities like viruses, which vary considerably and undergo continual evolutionary change (Van Regenmortel and Mahy 2004). It means that adaptations involving a single property can be regarded as taxonomically insignificant and that taxonomic decisions are always taken on the basis of a combination of viral properties.

"Thus, different virus species do not have sharp boundaries. Rather, they should be viewed as fuzzy sets with hazy boundaries" (Van Regenmortel and Mahy 2004). It is obvious that such fuzziness must have an impact on the construction of viral taxonomies. Some viruses have coevolved and cospeciated with their host organisms, and in these cases taxonomy will mirror phylogeny. The genomes of some viruses seem to have evolved modularly, however, acquiring genes or gene complexes intact from the genomes of other viruses or their hosts. If this has been the evolutionary pattern then, just as we saw in the evolution of prokaryotes, "trees deduced for one gene do not necessarily link viruses in the same way as trees deduced for a different gene" (Mayo and Pringle 1998: 655).

Species versus Ecological Clusters

The species concept does not merely arise from a human desire to impose order on nature, but is rather "an integral part of evolutionary theory" (Bock 2004a: 182). The genetic recombination that is involved in sexual reproduction is advantageous only if there are limits on the amount of genetic variation that can result from it. Species barriers protect a pool of genes, which have successfully evolved to survive in a particular ecological niche, from the risks of too much disadvantageous variation. In this sense, the biological species concept is only applicable to sexually reproducing organisms, because asexually reproducing organisms (or viruses) do not need to be protected from excessive genetic recombination (Bock 2004a: 178).

It is important to bear in mind, however, that there is not so much a dichotomy between sexual and asexual reproduction, as a continuum of reproductive gene exchange, ranging from almost every generation to almost never (Wilkins 2006a). Varying patterns of biological taxonomy can therefore be seen as the result of variations in the amount of gene exchange that occurs during reproduction.

At the 0 percent end of the gene exchange spectrum, individual lineages cannot recombine with any others, and any cohesion and clustering in either genotypes or phenotypes will be the product of ecological selection (Wilkins 2006a: 391): phenotypic adaptations are made to discontinuous environmental demands, with the result that clusters of organisms arise that are fit for a particular ecological niche (Bock 2004a: 184).

At the 50 percent end of the gene-exchange spectrum, however, ecological selection will be less important than compatibility of gene exchange. Obligate sexual species are defined in terms of their genetic isolation: their shared transmission of a unique gene pool. Individual lineages can recombine only with others in the species, and the species is maintained by the mechanisms of genetic, reproductive and ecological isolation. In this way, sexual species are protected from too great a number of disadvantageous gene combinations: each lineage can only be recombined with other lineages with which it is genetically compatible.

Somewhere between these two ends of the spectrum, in microbial lineages that occasionally share genes or gene fragments, "the greater the genetic distance between strains, the less likely it is that the transferred genes will be functional and useful in the receiving strain" (Wilkins 2006a: 396), so that clustering results from compatibility of gene exchange, as it does in sexual species, as well as from ecological selection.

The more recombination that is possible, the greater the degree of genetic rather than ecological selective pressure. Thus the patterns of biological taxonomy are affected by the ways in which organisms and their genes are reproduced. When organisms are sexually reproduced, clearer lines can be drawn around clusters than when they are asexually or virally reproduced. Tree-like phylogenies grow out of reproductive mechanisms that prevent once-separate genetic packages from leaking and their contents from remixing. Reticulate taxonomic patterns result, in contrast, from mechanisms that allow information to seep from one package into another. The identifying characteristics of any given species will have to be expressed increasingly polythetically, as the rate and degree of evolution increases.

Cultural Inheritance Mechanisms

Like biological information, cultural information can be reproduced via not one but many different mechanisms, including the biologically evolved psychological mechanisms of the human brain and the culturally evolved mechanisms of appliances like photocopiers, printers and radios. All that matters for the heritability of cultural information (as for any other kind of information) is that there must be receivers with the ability to make sense of it.

The receiver might be a photocopying machine in which an intense beam of light is reflected from a printed page onto a charged, photoconductive drum: the pattern of reflected light determines the pattern of negative and positive charge on the drum, which determines the pattern of negatively charged toner particles that are attracted to the drum, which in turn determines the pattern of toner that is attracted to a positively charged sheet of clean paper that is passed over the drum's surface (Meeker-O'Connell 2001). The photocopier, in other words, has been designed to interpret a particular pattern of black and white on a printed page as a negative pattern of toner particles on a photoconductive drum, and to express this pattern in an exact copy on a new sheet of paper. In effect, this machine acts as a receiver with the ability to make sense of incoming cultural information in a particular way. Its interpretative and expressive abilities are the product of human design.

When information is reproduced via a photocopier or printer, CD or the broadcast media, it is copied without input from any other information. We might say that there is only one cultural "parent," although it is worth clarifying what this means. Many writers, myself included, have described humans as having only two biological parents but many cultural parents. By this, we mean that although we receive our genes from only two people, we can receive our culture from any number of people, whether or not they are genetically related to us – and I think that the phrase "cultural parent" is a reasonable way to express this fact. Humans provide a mechanism for the reproduction of cultural lineages, just as hosts provide a mechanism for the reproduction of viral lineages – but although we might talk about Janet having caught a virus from Neil, this would not lead us to say that Neil is the viral parent of Janet. Rather, the parent in this context is the previous generation of virus: Neil's virus is the parent of Janet's virus. Similarly, in cultural reproduction, if we talked about Pauline having learned a piece of information from Jim, this does not, strictly speaking, make Jim the cultural parent of Pauline. Rather, the parent in this context is the previous generation of cultural information: Jim's copy of the information is the parent of Pauline's copy. Thus, although I don't think that it would be helpful to become unduly pedantic about the description of humans as cultural parents, it would be more accurate to focus on different generations of the information that they exchange.

So the question, here, is not whether information is ever received by one person from more than one other (which is trivially true), but to what extent cultural reproductive mechanisms allow different lineages of information to recombine.

Here, again, it is important not to be misled by too close an analogy with nature, where the mechanisms of meiosis ensure a precise, fifty-fifty division of parental genes. Because cultural reproduction does not take place via meiosis, there is no reason to assume that cultural recombination will involve the same equal division of material, or even that it will be restricted, necessarily, to two parental lineages. Metarepresentation is the mechanism that makes it possible for information to be combined from more than one cultural lineage. As we have seen, metarepresentation is a universal human capacity, which nonetheless varies, like height and eye colour, between individuals. Some people are more likely than others to question what they are told; to innovate rather than faithfully replicating; to analyse and synthesise incoming strands of information. It is also the case, of course, that each person is more likely to think in these ways about some cultural areas than about others, depending on the spread of his natural abilities, the skills that he has honed and the pattern of education that he has received. We are all capable both of transmitting information to other people, more or less unchanged from the version that we received, and of metarepresenting information with the result that novelties emerge, orthodoxies are challenged – or indeed that the information is verified and transmitted unchanged after all.

Because metarepresentation carries with it the potential for recombining any one piece of information with any other, can we see any reason why cultural information, like biological information, might benefit from segregation into discrete clusters? The answer is that if cultural information truly evolves, and if metarepresentation facilitates its recombination, then there is the same motivation for cultural clustering as there is for the emergence of biological species: the protection of cultural information, which has successfully evolved to survive in a particular cultural niche, from the risks of too much disadvantageous variation.

Cultural Phylogenies

We know from biology that phylogenetic trees provide a useful way of mapping the branched evolutionary history and relationships of bounded units of information, such as sets of genes that are passed on as fixed packages in one-off reproductive events. Phylogenetic trees cannot, though, chart the reticulate evolution of information that can escape from its original packaging. When this happens, new genotypes are created whose constituent parts do not all share the same evolutionary history. In some areas of biology, there are still more complications: it seems likely that viruses have had multiple origins; certainly, their movements in and out of host genomes, and between host species, will subject them to different selective pressures, adding further complexity to the patterns of viral evolution.

It seems obvious that culture's potential for ongoing, piecemeal reproduction will inevitably produce a reticulated rather than a branching map of descent, and that further complications will be introduced, in culture as in viral taxonomy, by the facts that many aspects of human culture are likely to have had multiple origins and that the selective pressures on cultural information will differ enormously from one receiver to another. It is perhaps surprising, then, to learn that several studies have successfully applied phylogenetic methods to cultural data, reconstructing the history of particular traits and identifying patterns of change; answering questions about whether groups of traits are related by descent and about the relationships between their selection and that of other traits (Mesoudi, Whiten and Laland 2006: 333). The success of cultural phylogenetic methods strongly supports the conclusion that metarepresentational cultural transmission, like biological sexual reproduction, produces an evolutionary structure that benefits from barriers to prevent excessive recombination.

When barriers prevent the intermingling of information, each receiver can only make sense of the information in its own pool, with the result that the evolutionary history of the pool will be the same as the evolutionary history of any information in it. Because information depends for its replication and expression on a receiver who can understand and act on it, its content makes no sense to someone who is not familiar with the language in which it is represented, and it cannot be expressed by anyone who does not have the relevant skills. Cultural units are therefore isolated from other units that depend for their replication and implementation on a different language, as well as from other units that depend for their expression on a different skill set.

I have emphasised the human capacity to acquire new languages, as a result of which we can make sense of a whole new range of information. If membership of a cultural species is defined for cultural information and its effects by a shared language, but the same human mind can acquire a whole range of such mechanisms, then it is tempting to deduce that information can be shared between virtually any given cultural species and any other. Indeed, cultural evolution has taken off precisely because of this unique human ability to extract information from one context and manipulate it in another, which brings with it the possibility of new species emerging from the convergence of old ones. We can see this, in practice, in the evolution of languages, of genres in literature and music, and in any other cultural area you care to mention. But we should be careful not to be carried along too far by the current of this argument. Cultural species barriers might shift and leak, varying the shape of their rather porous enclosures, but this does not undermine the whole concept of their existence. As demonstrated by the cultural phylogenies that researchers have produced, there are limits on the amount of movement and seepage that can take place. The following section will show how such barriers between different elements of culture can arise in practice, to the extent that we can meaningfully use the term *cultural species*.

The Mechanisms of Cultural Isolation

Information depends for its replication on a receiver with the ability to make sense of it. This means that if different lineages of information are to be recombined, then they depend on a receiver with the ability to make sense of them all. Situations that prevent separate strands of information from reaching the same, appropriately equipped human will therefore be effective mechanisms of cultural isolation. Just as the genetic barriers between some organisms are geographical – they simply never meet – so some areas of culture are geographically or historically isolated from each other: there is never any overlap between the humans who originate or encounter them. So long as there is no overlap between the people who read one set of books and the people who read another set, for example, the information in those books will remain isolated from each other.

As well as isolation mechanisms that are extreme enough to ensure that no human ever encounters certain combinations of information, there are also situations in which the combinations are potentially available, but the effects of one informational package make it unlikely that the same human would acquire the other. Some cultural information has, as one of the effects of its acquisition, the exclusion of some other information from its receivers' consideration. For instance, you won't have the chance to learn the rules of cricket if you have already accepted the claim that cricket is boring. Similarly, if you reject a particular genre of television show, book or music, then the content of that genre will be isolated from any other information that you do acquire. The point about these examples is not the self-evident truth that every one of us will be led towards some cultural possibilities and away from others by the information that we have acquired in the past. The point is, rather, that if *everyone* who accepts a particular portion of information tends, as a result, to avoid a certain other cultural area, then the population-level consequence will be that those two cultural areas will be effectively isolated from each other.

This can also happen as a result of such unfamiliarity with a particular area of culture that it makes no impact on our minds, even when we do encounter it. People who do not subscribe to the cultural practice of looking fashionable, for example, will probably not even notice whether other people are wearing the latest fashions or designer labels, and will take no information from the fashion pages of magazines even if they should glance at them. There can be similar barriers between the cultures of different nations. If you have no experience of British society, then you cannot decide whether hereditary peers, private education or the disestablishment of the Church of England might be good or bad things for Britain. You can't understand what the twenty-two people on a football pitch were doing a minute ago, why the man in black has just blown his whistle and why the other people have stopped what they were doing, if you don't have certain information about game playing in general, football in particular, and more specifically the off-side rule. This depth of unfamiliarity with a subject area is a very effective reproductive barrier between its content and the information that you have acquired, because it means that you cannot act as a receiver of any information that you might encounter in that area.

Academia provides many examples of this phenomenon. It is obvious that if you have never studied the philosophy of logic or language then you cannot know which theory of truth is the best; if you have never studied particle physics then it will be impossible for you to choose between rival theories in quantum mechanics; if you have never studied history then you will be unable to decide between alternative historiographical approaches. What might be less obvious is that such unfamiliarity with a subject area means that you cannot even act as a receiver of much of the information that it encompasses. If, for example, you happened to read something that was based on a question-begging assumption that one particular theory, out of several rivals, is correct, then you would not recognise the underlying assumption because you don't know anything about the surrounding debate. Once again, the important point here is to be found at the population level: if everyone who acquires one portion of cultural information is blind to the presence of another portion when they encounter it, then these two cultural areas will be effectively isolated from each other. In reality, of course, this will often be the case. Although it is theoretically possible for one person to become familiar with any combination of cultural areas, in practice there are many possible combinations with which nobody is familiar.

This is as true of different language combinations as it is of different sets of informational content – and indeed the two are intimately connected. Exclusion from a jargon-laden specialist language will often reflect a more general unfamiliarity with the concepts expressed by that jargon: a lack of training in a particular cultural area. There is a peculiarly disorienting feeling that comes from a first encounter with the mutually dependent vocabulary of a new subject, especially if you have recourse to a specialist dictionary. Turning from one page to another, you are led through an entwined series of conceptual loops from which it is perfectly possible to emerge none the wiser. "Fallacy: an invalid argument with the appearance of validity." But what is an argument and how could it have the appearance of validity? "Argument: a complex consisting of a set of propositions (called its premises) and a proposition (called its conclusion)." "Validity: Any argument is valid if and only if the set consisting of its premises and the negation of its conclusion is inconsistent."

This is an example created by opening a dictionary of philosophy (Honderich 1995) at a random page, picking a word and quoting only a tiny fraction of the information provided about it, and then turning to the definitions of a couple of words within that information and quoting only a little about them, too. I could go on to quote from the definitions of other words crucial to our understanding of the foregoing definitions – *set, proposition, negation, consistent* – and thus increase the number and entanglement of conceptual strands. Readers unfamiliar with logical philosophy, who felt baffled by the jargon within those brief quotations, may not even have realised that they did not understand one of the key phrases included: "if and only if" has a precise, technical meaning within logic.[1]

What such examples show is that it is possible for someone to speak English perfectly well, yet still have no understanding of the specialized languages that are used in particular areas of culture. Although expressed within the context of English syntax and familiar words, an excess of unfamiliar vocabulary can preclude understanding as effectively as an unfamiliar foreign language. These variations can arise in academic circles, regional uses of a shared language and specialist or minority interest areas like music or fashion.

More fundamentally still, of course, there are natural language barriers. You cannot understand what someone says in an unfamiliar language or what is written in that language. The same applies to artefactual languages: you cannot learn your times tables until you have mastered some basic mathematical notation; if you have learned to play the guitar with the aid of tablature, then you cannot play a new piece of music that is represented in staff notation. Here again, the key point is that, although it is theoretically possible for any combination of languages to be acquired by an individual human, in practice there will be many possible combinations that nobody in the world has, as it happens, acquired. Any information that depends for its representation on natural or artefactual language F will be isolated from any information that depends on language G, if there is nobody in the world who knows both F and G.

Even when a human receiver is capable of making sense of the two separate strands of information that she encounters, there will be times when no productive metarepresentation takes place: the receiver is appropriately equipped to understand and act on both strands of information, separately, but for some reason is unable to make links between the two. The reason for this might be very simple: this person is not very

---

[1] "*p* if and only if *q*" is an abbreviated way of asserting both "if *p* then *q*" and "if *q* then *p*."

good at metarepresenting, and although well able to retain and pass on information, she will not often synthesise or evaluate different informational strands. More generally, even people who are highly metarepresentational in some cultural areas will be unable to reflect, make links and be original in other learning areas. Whatever the reason, so long as the people who acquire both sets of information are unable to make links between them, these two informational strands will remain isolated from each other.

Even when the receivers involved are very able to metarepresent the information that they acquire, it will not be possible for them to make links between certain cultural areas. In particular, this will happen when concepts that are represented in one language cannot be translated into another language. The content of cultural languages – the concepts that they represent – coevolve with the languages themselves, and there will often come a point at which the content of one language cannot be represented in another. How, for example, would you set about representing a piece of music using the conventions of engineering drawing, or a Laplace operator on a clock face? When an artefactual language has evolved to represent the content of one particular cultural area, it will be ill suited to represent the content of any other.

Many concepts that were dependent on a particular artefactual language for their emergence can nonetheless be expressed, albeit clumsily, in natural language, and this might seem to suggest that we can use natural language as a bridge between incommensurable artefactual languages. The problem with this suggestion is that there will rarely be any potential links between the content of one specialist artefactual representation and another: even if both are expressed, longhand, in natural language, no connections will emerge. There are no more fruitful connections to be made between the phrases "the legal right to control the production and selling of an original literary, artistic or musical work" and "when one variable, which we'll call t, changes, another variable, which we'll call x, changes thirty-eight times faster," than there are between the symbolic representations of these phrases: © and $dx/dt = 38$.

When languages represent such distant semantic fields, the metarepresentational attempt to compare and contrast their content, seeking links and new perspectives, will be to no avail, even when we can use the same language to represent it all. I can use natural language to talk about the techniques of cake baking, the fuel efficiency of my vehicle and the rules of Newmarket: none of these is an area in which I need to resort to any specialist artefactual language for more effective representation of the sorts of thing that I might say about them in everyday conversation. Anybody who is conversant in English will be able to grasp what I am saying about all of these subjects, even if she has not herself baked a cake, calculated a vehicle's fuel efficiency or seen a pack of playing cards – but this does not mean to say that there will be any fruitful grounds for comparison or recombination of the information about these separate subjects. We can think and talk about potential links between them, but ultimately there would be little to be gained from so doing. It is certainly unlikely that any connections that could be made among fuel efficiency, card games and baking would be successful enough to be replicated.

It is clear, then, that cultural barriers do exist and that they are largely language based. Cultural information that has evolved to survive in a particular cultural niche is protected from the risks of excessive variation by its basis in a particular cultural language. It depends for its reception and transmission on receivers who know its cultural language, and in order to be recombined with information from another cultural area it would need a receiver who knows both cultural languages. Even then, the fact that languages tend to coevolve with their content often means that recombination will not be possible.

Cultural areas are effectively isolated from each other when the language in which one is represented cannot also represent the content of the other. A receiver who acquires information from both areas is unable to use metarepresentation to bridge the gap between the two, because there is no common ground between them. Concepts from one cannot, therefore, be recombined with concepts from the other, either because there is no way of representing them in the other language or because the content of each is incommensurable with the other. Cultural species barriers are also maintained by natural restrictions on the human capacity for learning. Although in theory each one of us could acquire any combination of linguistic and practical skills, in reality there are practical limits on the number of skills and languages (both natural and artefactual) that each person can acquire. In practice, too, there is a limit to how much the human brain can keep current at any one time: unused skills attenuate, and informational memory can fade. None of this erects a theoretically impenetrable barrier between two fundamentally commensurable areas of culture, but what it does mean is that in practice cultural species barriers are often less likely to be breached than they might seem to be in theory.

Carving Culture at Its Joints

Elliott Sober (1993: 157) has suggested three conditions against which any proposed biological taxonomy should be evaluated: it should be clear, making obvious where we should draw the lines; it should be theoretically motivated, proposing groupings that are biologically significant; and it should be conservative, proposing groupings that fit the use that biologists have actually made of the species concept. So although it is perfectly possible to draw clear lines around organisms on the basis of their size or colour, for example, this would be theoretically pointless and totally at odds to the usual practice of biologists.

If any proposed cultural taxonomy is to be taken seriously, then it is important that it should share these advantages with its biological counterpart: it should enable us to draw fairly clear lines around cultural clusters, in a way that is in keeping with current understandings of culture and meets the theoretical demands that we have put on it. How does a language-based cultural species concept, which is based on the concept of a receiver who is able to make sense of concepts that have coevolved with particular cultural languages, fare against Sober's criteria?

Certainly, a language-based cultural species concept enables us to delineate cultural clusters or species. Artefacts, behaviours and sets of ideas are grouped together by common inheritance mechanisms: their shared accessibility by a particular cultural language, with its associated cultural content. These items of furniture are all tables because you can use and select between them all with the help of the same set of background information and concepts. In this sense, they have more in common with each other than this wooden table does with that wooden chair, even though some of these tables are made from glass and metal; and they have more in common with each other than the black ones do with black cars or than the child-sized ones do with child-sized cutlery. A language-based cultural species concept enables us to answer questions about why one cultural artefact should be classified in the same species as another but in a different species from a third; and the answers that it gives are both theoretically useful and in keeping with our folk understanding of culture. The taxonomic lines that it draws around clusters of cultural artefacts and behaviours have a good fit with both our everyday understanding of the differences between artefacts or behaviours and the distinctions that are made by other cultural disciplines. A language-based species concept requires us neither to reconceptualize our intuitive understanding of culture nor to dismiss the findings of disciplines like sociology and archaeology, psychology and anthropology.

In practice, of course, we will recognise groups of artefacts or behaviours not on theoretical considerations about language use but rather on (often-polythetic) morphological criteria. They will be distinguished from other such groups by their occupation of a particular cultural niche, and possibly by their cultural isolation. This means that we can recognise all of these artefacts as dining tables, for example, because they all have a flat horizontal surface supported by one or more legs (morphological criteria). We can distinguish them from other, similar artefacts by looking at both their morphological features and the cultural niche that they occupy. One artefact with four legs and a flat horizontal surface might also have drawers underneath and three enclosed sides, making it more suitable for use as a desk than as a dining table; another might have a baize-covered top, more suitable for playing billiards than for dining. One artefact with a small, high, flat horizontal surface supported by a single leg might be made of stylish metal and glass, making it suitable for use in a pub or bar; another artefact with a similar design, made from stained wood, is more suitable for outdoor use as a bird table. Conversely, we place all of these artefacts in the "dining table" cluster, even though one is made of oak, one of marble and another of wrought iron; one has four legs and another has six; one is two-feet square and another can seat twenty-four people. Artefacts do not have to share identical morphology to be placed in the same cultural species: there will be a (possibly polythetic) set of morphological criteria by which we *recognise* an artefact as a member of a particular cultural cluster, but its cultural niche and possible cultural isolation is what defines the species. As in biology, the fact "that most species taxa are recognised on the basis of morphological criteria is immaterial to the definition of the species concept" (Bock 2004a: 179), which is founded on broadly linguistic criteria, be the language genetic or cultural.

As for the theoretical motivation for a cultural species concept, what truth about cultural classification are we seeking? We may feel able to dismiss the suggestion of classifying man-made items according to colour, just as colour is not a useful category in biology, but it is not so obvious how we should categorise that child-sized table. Should it be placed in the nursery section of a furniture store or in the furniture section of a childcare store?

Intuitively, I would see no contradiction in a furniture store's selling a child-sized table any more than in a childcare store's selling one. Does this mean that these are arbitrary decisions, or is there some theoretically well-motivated criterion against which we can judge each option? The advantage of the cultural species concept is that it allows us to resolve these intuitive dilemmas. If I had all of the information and skills needed to utilise a table, would this enable me to utilise a child-sized table? Yes, because the concept of a table embraces the possibility of different sizes. If I had the concept of child-sized objects, would this information enable me to utilise a child-sized table? No, because the table-related information would still be missing: the concept of child-sized versions of artefacts does not embrace information about every possible artefact that can be miniaturised in this way. So although I still wouldn't argue with a childcare store's decision to sell child-sized tables, the cultural species concept enables me to classify this item for theoretical purposes.

This is not to say that hybrid cultural species are not possible. To pick trivial examples, it is possible to manufacture a pen that incorporates a calculator, or a baked-bean tin that doubles as a secret safe for small valuables. In order to utilise or reproduce hybrid artefacts like these, you would need input from not one cultural cluster but two. In fact, it is arguable that the cultural barriers around such hybrids are even stronger than those around the parent species, because in order to understand and implement the hybrids you need both of the relevant portions of information. You might even need the scaffolding of a new cultural language.

Unlike in biology (or at least in eukaryotic biology), where "there are no awkward intermediates" as long as "we stay above the level of the species" and study only "animals in any given time slice" (Dawkins 1986: 260), in culture genuine intermediates do exist. We carve as close to culture's joints when we put the pen-calculator in the calculator section of the shop as when we put it in the pen section. But this is not to say that cultural taxonomy is arbitrary: that we might as well put the pen-calculator in a seven-and-a-quarter-inch-long section of the shop, or on the same shelf as the other artefacts that were manufactured in July. We can do this, of course, just as we are free to classify the natural world according to leg number or time of birth. All of these strategies would be theoretically pointless, however. Just as in biology we seek to base our taxonomies on a theoretically well-motivated account of reality – the way in which organisms are biologically closest to each other – so in culture we should seek to base our taxonomies on a theoretically well-motivated account of reality: the way in which artefacts or behaviours are culturally closest to each other. To understand and use a pen or a calculator, it is not necessary to know its date of manufacture or its precise length. The fact that the resulting cultural taxonomy is not always a nested hierarchy does not mean that it is not based in reality: that it does not carve culture at its joints.

As in biology, patterns of cultural taxonomy will not always be straight-forwardly phylogenetic or reticulate: more complex taxonomic patterns are possible, depending on the representational and ecological pressures at play. This is the case, for example, in areas of material culture such as Turkmen textiles, New Guinean material culture, East African marriage patterns and Neolithic pottery found in Germany (Tehrani and Collard 2002: 453–4). In all of these diverse cultural areas, research has revealed patterns of reticulate, ethnogenetic taxonomies overlaying, to a greater or lesser extent, a more dominant phylogenetic cultural pattern. "Thus, the results of the quantitative studies of cultural evolution that have been reported in recent years argue strongly in favour of case-by-case empirical assessments of the relative contributions of phylogenesis and ethnogene-sis to cultural evolution" (Tehrani and Collard 2002: 454): although geo-graphical proximity can have a cross-cultural impact in some cases, other cultural factors like matrimonial patterns, historical hostilities between neighbours and the nature of the cultural content involved will also play their part.

Peter Jordan and Stephen Shennan (2002), for example, have looked at the taxonomic patterns of basketry traditions in different groups of Californian Native Americans. They explored the relative influences of linguistic, geographical and ecological factors on these traditions, and found evidence of some cultural transmission across local linguistic boundaries. The authors conclude that there is no clearly defined, close relationship among languages, material cultures and ethnic identities, and they discuss the possibility of different rates of cumulative evolution-ary change in local linguistic and material traditions (2002: 72). Indeed, there is no reason to suppose that different cultural areas, which depend on different systems of representation, should evolve at identical rates. The content of each will coevolve with its own language and medium. The factors that affect cultural evolution in tribal basketry traditions will include the role that basketry plays in tribal identities; the level of trade or matrimonial contacts between tribes; the materials available locally; the ways in which basketry skills are learnt, and by whom; and so on. Only some of these factors will also affect cultural evolution in tribal natural languages, and therefore we should not expect to see an exact parallel between the two.

The structures and designs of traditional Turkmen weaving, on the other hand, show a different taxonomic pattern. These were skills passed on by imitation and verbal instruction from mother to daughter and reproduced from memory (Tehrani and Collard 2002: 457). They were thus acquired over many years, from early childhood, and this raises the possibility of effective species barriers between the weaving styles of different tribes, as in practical terms it would have been very difficult for one tribe to adopt the structures or designs of even a neighbouring tribe. Moreover, Tehrani and Collard (2002: 458) suggest that Turkmen carpet designs might have provided a language-like tribal identifier, especially given the dependence of weaving skills on being acquired over many years, from early childhood, in a particular tribal context. If this is the case, then the predominantly tree-like phylogeny of Turkmen weaving styles is unsurprising: like each tribe's natural language, its weaving style has culturally evolved as much for the exclusion of strangers as it has for the identification and inclusion of tribe members.

In culture, as in nature, taxonomy is the product of both ecological pressures and the varying reproductive mechanisms to which information is subject. In both, clustering results from divisive ecological pressures and from the protection of species barriers against too much recombinative variation. Just as there is a spectrum in biology, from obligate asexual transmission to obligate sexual transmission of genetic information, so there is a cultural spectrum from the straightforward replication of cultural information, through borrowing and imposition, to metarepresentation. Where cultural information depends for its transmission on the prior transmission of a whole new linguistic package, we should expect to see more tree-like phylogenies. Where transmission is possible on the basis of an existing linguistic package, we should expect to see more reticulate phylogenies, with less equivalence between the evolutionary histories of packages and their components. In these ways, patterns of cultural taxonomy are shaped by the ways in which cultural information is transmitted, just as the patterns of biological taxonomy are shaped by the ways in which genetic information is transmitted between generations.

14

Conclusion

A Representational Understanding of Cultural Evolution

Human culture is extraordinarily diverse. Any theory that purports to explain the origin of its variety and complexity must be able to account for all of it, from atheism to zabaglione. Any credible theory of culture's origin will be expected to answer questions about whether animals have culture; about the relationship between human culture and human biology; about the role of language in facilitating human culture. It will need to be able to account for the nature of the changes in human culture over the millennia, for the mechanisms by which its complexity is maintained and transmitted and for the patterns of cultural diversity that have emerged. The explanations that it provides for observed cultural traits like particular artefacts or their features, particular behaviours or their constituents, must be both grounded in the empirical evidence and precise enough to invite challenge and refutation where appropriate. A theory that can provide a glib explanation of any observation that you care to mention has, of course, little chance of providing an adequate explanation of anything. Cultural explanations must grow out of cultural observations, and cultural predictions must be falsifiable.

Cultural Evolution

For the majority of organisms, their behaviour is as automatic a developmental product of their genes as is their physical form. In a large minority of organisms, however, there is an experiential buffer between their genotype and some aspects of their behaviour. In these species, one developmental product of their genetic information is the capacity for individuals to learn from their experiences. In a minority of these species, members can also learn from one another. In only one species, so far as we know, can members learn from what they have learned.

The deeper the buffer between genes and their behavioural expression, the more we have to look to an organism's past experience for a proper explanation of its current behaviour. Human cultural behaviour is so cushioned from the impact of our genes that biological explanations are of limited assistance to our understanding. Although it is true that human culture could not have emerged in a species that had not been prepared by the processes of biological evolution, it is also true that the ways in which our culture emerged, and the multiplicity of its current structures, cannot be explained by reference to the developmental expression of genes alone. In this book, I have explored and defended the theory that human culture is best explained as the product of evolutionary processes in culture itself.

Heritable Information in Nature

I have argued that culture is the product of gradual changes in heritable cultural information and have shown how this claim is founded on a robust theory of information and its inheritance. The essence of this theory is that information cannot exist in isolation but always relies on a receiver that can interpret and react appropriately to it. Such a theory can contribute both to our understanding of cultural processes and to the resolution of certain controversies within biology.

Current biological debates set developmental systems theory (DST) against the idea that genes are causally privileged in organisms' development. These debates are founded on the assumption that DNA is *either* a genetic blueprint *or* one source of information among the many that contribute to developmental systems. If, however, we see heritable information as a resource that is transmitted between prepared receivers, then we can accept both that many different resources contribute to developmental systems and that the traditional distinction between phenotypes (on which selection acts) and genotypes (whose differential survival produces evolution) does matter.

First, this account of heritable information reveals that DST is right to emphasise that naked genes alone will never be enough, because information always needs the right kind of receiver: a receiver that can discretely represent variations in the source of information and respond appropriately to them. It also supports DST's claim that genes do not, after all, play a causally privileged role in development, if this privileging means that their information has been selected to *prescribe* a phenotype, whereas environmental conditions are merely predictive of phenotypic features. In fact, all information is prescriptive, in the sense that it prescribes the reactions of any receiver that "knows" how to respond appropriately to its variations. Night length prescribes a photoperiodic plant's flowering in as real a sense as the plant's DNA prescribes the plant's response to variations in night length.

Secondly, however, this account of heritable information confirms the sense in which DNA does play a privileged role in biological evolution. The cellular machinery for DNA's transcription and translation is the receiver without which genetic information would effectively cease to exist, and in this sense it is true that DNA cannot act in isolation. But DNA's privileged role in evolution does not result from any alleged capacity to exist and act in isolation. It results, rather, from DNA's spectacular capacity to ensure the persistent heredity of its variations. To the right receiver, DNA transmits information not only about an organism and its behaviour but also about how to copy the information that it carries (viral DNA doesn't even bother with the organism). Phenotypic variations can also result from variations in external resources, such as the environment and the behaviour or bodily products of other organisms, but it is the inheritance of genetic information which ensures that each biological generation can act as a receiver of the information in those other resources. If we are giving a complete explanation of how a phenotypic outcome was produced, then it is true that we need to invoke a whole range of causal factors; but the genetic factor will often be the one that explains why the outcome was what it was.

Further phenotypic variations can result from variations in the cellular machinery, but it is unclear to what extent these epigenetic variations persist through the generations. If they can sometimes persist through multiple generations, then epigenetic inheritance will provide an additional mechanism for biological evolution, but the jury appears, at the time of writing, to be undecided on this question. What we do know is that the mechanisms of genetic inheritance ensure the persistent heredity of variations in DNA.

The Mechanisms of Cultural Inheritance

This book's theory of heritable information enables us to understand why it is so important for evolution, in any sphere, that the inheritance mechanisms on which it relies are able to ensure that each generation receives not only information but also its means of interpretation and transmission. In culture, natural and artefactual languages are the inheritance mechanisms that play this role.

Human receivers are biologically prepared to receive information not only from but also about any representational system that they encounter. This matters because, for one person to receive cultural information from another, both must understand the way in which the information is represented: the first to act as a source of information and the second to act as a receiver. The receiver must discretely represent the information that the source provides, in the same way that the source discretizes it. When we speak to another person in our shared natural language, this objective is achieved. The evolution of natural language therefore provided the foundation for cultural evolution, because it ensured a steady supply of receivers who all discretely represented cultural information in the same way. It is not a representationally perfect system, and it sometimes relies for clarification on extralinguistic techniques such as gestures, facial expressions or Gricean assumptions about relevance and cooperation. It was certainly sufficient, however, for the early evolution of cultural information.

The view of natural language as an inheritance mechanism for cultural information enables us to understand how evolution became possible in human culture in a way that it never has done in other species. What it cannot fully explain is the galloping pace of cultural evolution in latter centuries, and how its accumulated complexity and diversity are supported. Cultural information has long outgrown the collective capacity of human brains and natural language, and much of it now relies on the inheritance mechanisms of artefactual languages for its preservation and replication. It does not matter how clever an individual human might be: if he cannot convey his thoughts to other people, then his innovations cannot be subject to cultural evolution; and if his thoughts are complex enough to need the scaffolding of artefactual representation, then he will not be able to convey them to anyone who does not understand the system in which he represents them. The correlation between the cumulative evolution of complex modern forms of cultural content and the evolution of its artefactual systems of representation is thus no coincidence. Artefactual languages have all sorts of representational advantages over natural language: they provide not only cognitive tools like scaffolding and swap space, which facilitate the emergence of specialised cultural areas, but also the means of transmitting the newly evolved content from one individual to another in a form that is accurate, persistent and mutually comprehensible.

Metarepresentation and Nonhuman Culture

As well as facilitating communication and representation, languages also shape the cognition of those who learn them. Receivers must learn to discretize information in the same way as the language in which they receive it, and this is bound to have an impact on the ways in which they conceptualise that information. The great virtue of the human language capacity, from the perspective of cultural evolution, is that it enables us to escape those cognitive limitations by reflecting on the information that we acquire and on the languages in which we acquire it, learning new languages and re-representing the information in new ways. This universal capacity for metarepresentation is what has powered cultural evolution in humans. What about culture in other species?

Nobody thinks that humans are the only creatures with mental representations, or the only creatures who can copy one another's behaviour, or the only creatures with any semblance of culture. Where humans do differ most strikingly from other creatures, however, is in the extent to which we are capable of acquiring information and the extent to which we are prone to sharing it. The motivation to share information with conspecifics is dependent not only on membership of an essentially cooperative species but also on the metarepresentational knowledge that one has information worth sharing. The ability to acquire information from conspecifics is dependent on the capacity to discretize information in the same form as that in which its originator is offering it: only once a species has a steady supply of members that share the same method of representing information can that information be exchanged. And only once a species is able to reflect on the information that it is sharing, and on its method of representation, can there be evolution in both the information itself and its representational systems. No other species, so far as we can tell, has either a shared representational system or the metarepresentational ability to develop one, making humans unique in our ability to exchange cultural information with sufficiently persistent and differential heredity to support cultural evolution.

Social versus Functional Links

The human tendency for cooperation has been crucial for cultural evolution, but natural and artefactual languages support different types of cooperation. The cultural evolution of natural language was accelerated by the biological advantages of enhanced communication among members of a cooperative species. As a consequence, natural languages are important markers of social identity, and they exclude outsiders from a social group as effectively as they define which individuals count as insiders and facilitate communication between them. Artefactual languages, on the other hand, evolved under adaptive pressure for more effective representation, and one of their representational advantages over natural language is that they can be detached from their human originators. This enables information not only to be disseminated over much greater expanses of time and space than the content of speech but also to shed the social associations of its human originators. If I speak to you in an accent that you have been brought up to regard as uneducated (and by common implication unintelligent), then you might not take seriously what I have to say. If I type the same information and send it to you on a piece of paper, then you may be more likely to focus on the content than on its originator. When philosophers talk about the fallacy of ad hominem arguments, they refer to the temptation to reject an argument, or conversely to accept an argument's validity, simply because of who has made it. Although logically this is indeed a fallacy, in our daily lives it makes perfect social sense to give more weight to the words of our friends than to those of strangers. What artefactual languages make possible, with their focus on representational precision rather than communicative ease, is the isolation of informational content from social context. They make it easier for us to put our relationships to one side in order to do business with one another. They facilitate, in other words, functional links between groups of otherwise-unrelated individuals.

It is important to recognise that humans have an innate capacity for both social and functional cooperation, which is supported rather than created by the acquisition and shared use of natural and artefactual languages. Individuals' varying capacities for social and functional cooperation are seen in the different uses to which languages are put by different types of people. A high-functioning individual with Asperger syndrome, for example, has a good grasp of natural language but uses it almost exclusively for representational purposes, creating and maintaining functional rather than social links with other people. A high-functioning individual with borderline personality disorder, conversely, will use natural language almost exclusively as a tool for social cooperation – but it is important to remember in this context that not all social links are positive. This sort of person is extremely competent at making social links – the nature of which, ironically, makes it almost impossible for her to work productively with others: this is the person, to give a trivial example, who is prone to complaining that it wasn't what somebody said that upset her but the way in which it was said. Readers who are familiar with the "extreme male brain" theory of autism (see, e.g., Baron-Cohen 2003) will notice its links to my ideas here. Is borderline personality disorder perhaps the manifestation of an extreme female brain? Of course, my thoughts on this subject are at this stage largely speculative, but in previous chapters I have shown how the distinction between functional and social cooperation can help to explain a host of cultural phenomena, from the failure of televotes to match the result that judges might have produced in televised competitions, to apparent anomalies in societal uses of money.

Cultural Taxonomy

When humans acquire cultural information, they do not only store it but also tend to act on it. Much cultural information relies for its expression on individually learned skills like campanology, knitting or bowling, as well as on more general human capacities like manual dexterity, physical coordination or visual acuity. Culture is the product of interactions between this variety of cultural agents and the variety of cultural information that they acquire. It takes the form of identifiable units, whether artefactual or behavioural, which are assembled under the guidance of ecological and representational pressures into recognisable cultural clusters.

The view of cultural evolution as the product of heritable variations in cultural information, facilitated by the mechanisms of cultural languages, enables us to understand the patterns of cultural taxonomy that empirical research reveals. A crucial factor in that explanation is the spectrum of metarepresentational abilities across human populations, which determines the extent to which cultural information is straightforwardly copied and the extent to which innovations arise from recombinations. Variations in the human capacity for metarepresentation have had the same impact on patterns of cultural taxonomy as variations in the mechanisms of reproduction have had in biology. A human receiver will need to understand all of the languages in use, in order to lift information from its various sources and compare it in a new context. Species barriers, in culture as in nature, ensure that the potential for recombination does not squander the hard-earned benefits of earlier evolutionary processes.

A variety of phylogenetic analyses have provided an existence proof of cultural species barriers in both natural languages and material culture, although there are also areas of culture in which units form reticulate clusters of polythetic classes. The patterns that emerge in any particular cultural area will depend on the degree to which cultural information can be accessed by people who do not share the languages in which it is represented. Monolinguals simply cannot access any information that is represented in a natural language unrelated to their own, for example, and for this reason there are fairly impermeable barriers between natural languages, giving rise to phylogenetic patterns of linguistic taxonomy. More complex taxonomic patterns have been found in areas of material culture such as Turkmen textiles (Tehrani and Collard 2002) and Californian Native American basketry traditions (Jordan and Shennan 2002), where different representational and ecological factors are involved.

The Origin of Culture

I have presented evidence from diverse areas of research for the view of culture that this book supports, and have demonstrated its explanatory value in a variety of ways. It has answered questions about whether animals have culture; about the relationship between human culture and human biology; about the role of language in facilitating human culture; about the mechanisms by which changes in human culture over the millennia have been maintained and transmitted; about the patterns of cultural diversity that have emerged. These cultural explanations have grown out of cultural observations, grounded in the empirical evidence. Unfortunately, there is no greater likelihood in culture than there is in nature that evolutionary theory will lead us to make successful predictions, for in any evolutionary sphere it is only with hindsight, based on the performance of several generations, that it is possible to say whether a given variation has been advantageous or disadvantageous (Professor L. S. Shashidhara, personal communication). Nonetheless, in any proposed evolutionary sphere it is possible to offer explanations, and make predictions, that are precise enough to invite challenge and refutation where appropriate.

For instance, the claim that evolution can explain a particular type of complexity and diversity predicts the existence of a mechanism that can ensure the persistent heredity of variations in that diverse complexity: a representational theory of heritable variation enables us to test this prediction in any evolutionary sphere that is suggested – and in the case of culture, as in the case of nature, the prediction has been confirmed. Evolutionary theory predicts that, because fitness is relative, the success or failure of any given variety will be determined more by adaptive pressures than by absolute values: the biggest factor in information's success will be its adaptation to gain and retain the resources for which it is competing. In culture the resource is human attention, and evidence from urban myths, advertising and politics provides ample confirmation of this prediction. Evolutionary theory predicts that, over time, some variations will survive at the expense of others, and in both nature and culture we find evidence of extinct varieties, as revealed by palaeontology, archaeology and history.

An evolutionary claim that a particular modern variety is descended from an extinct historical variety predicts the possibility of finding a missing link between the two, and in culture as in nature such predictions have been made. Several specific predictions stem, for example, from the intuition that writing evolved as an artefactual system for the representation of oral language: this hypothesis predicts that early writing systems should phonetically and serially code the same broad semantic field as the spoken language (Damerow 1999). These predictions were falsified by discoveries that proto-cuneiform scripts do not code phonetically, that they organize information hierarchically rather than serially and that their semantic field is restricted to economic information.

In addition to explanatory and predictive value, the third test of any convincing theory of culture must be its practical contribution to problems and research programmes in the real cultural world. Cultural evolutionary theory, of the type that I have defended in this book, has been used to inform research and practice in a surprisingly wide variety of cultural areas. For instance, a research team at the BBC World Service Trust drew heavily on cultural evolutionary theory to inform the creation of their media outputs for a malaria communication campaign in Cambodia (Khosla 2008). This approach helped the team to prioritise and focus their messages against the background of a complex and contradictory set of malaria beliefs and practices in their Cambodian audiences. A representational view of cultural evolution provides the theoretical underpinning to Price's (2008) examination of the impact of workspaces on organisational culture. The international collaborative project, Visual Exploration of Cultural Design in Style, which is investigating the design style features of South Korean and Spanish cultural artefacts, has productively related the concept of cultural DNA to its methodology (Professor

Ji-Hyun Lee, personal communication). In addition, I am aware of doctoral and master's theses that have been completed in recent years, which apply a representational theory of cultural evolution to cultural areas as diverse as the professional conservatism of teachers in India; aspects of the United Kingdom's National Health Service; sex-ratio dynamics in China; font design and typography. There is some justification, it seems, for looking forward to the confirmation of this theory's "explanatory and exploratory power" (Price 2008: 54).

A representational theory of cultural evolution explains human culture as the product of evolving information. This information is discretized by human cognitive processes; it is preserved and transmitted in the form of shared cultural languages; it is expressed in artefactual or behavioural units by the cognitive and physiological activities of human agents; it is differentially replicated as a result of factors including human cognitive biases, the existing cultural context and the language in which it is represented; and as a result of this differential replication, these cultural units form clusters that, at least in some cases, can usefully be called cultural species. In this way, the origin of culture can be explained by means of natural and artefactual language.

Appendix

What about Memetics?

The essence of my theory is that culture, like nature, is the product of evolved information. Every other aspect of the way in which I explain human culture flows from the wellspring of that statement. Many readers will know that in the past (Distin 2005) I have made use of a handy conceptual tool that Richard Dawkins (1976) has given cultural theorists and referred to units of cultural information as memes. I am aware, however, that this terminology can so distract those readers who are in the habit of dismissing memetics out of hand, that they are unable to hear what else I am saying. Although a burgeoning optimism about cultural evolution is detectable across a variety of disciplines, memetics has been widely criticised and perhaps even more widely misapplied to a variety of irrelevant subjects. The World Wide Web in particular is full of pages and blogs that use the term *meme* with varying degrees of vagueness, often not bothering to define it at all but simply stretching it to fit whichever space has opened up in the writer's vocabulary. The intellectual credibility of memetics is diminished every time a meme-related term is hijacked in this way and its sense redirected to the latest cultural phenomenon to have caught the eye. This is one reason why memetic language can provoke a hostile reaction, and I would urge its critics not to be misled by the manifold ways in which it has been misused, to think that memetics itself is as vacuous as so many of its applications have been.

Even those cultural theorists who are able to see past its misapplications to the more serious accounts of memetics, however, can have grave reservations about its validity. These stem in part from hostility to the particular version of biological evolution in which memetics has its roots, and in part from concerns that theories of cultural evolution can be constructed perfectly well without recourse to memes; that there is, in any case, no evidence for memes' actual existence; and that memetics has no real explanatory value. In my opinion, none of these concerns is well founded: a theory of cultural evolution can be valid even when it includes memetic elements, and our understanding of human culture is not only unified but also enriched by such an account. More specifically, my theory is that human culture is built by human agents on the basis of cultural information, which they are able to create and acquire by virtue of cognitive mechanisms that discretize cultural information in ways that match the discretizing methods of the cultural languages in which this information is shared – and it will be obvious to anyone who has a passing acquaintance with memetics that any statement about discrete units of cultural information could be rephrased as a statement about memes. Because information must be discretely represented if we are to receive it, there is no real reason why we should not use the *m*-word when talking about discrete units of cultural information.

Nevertheless, there are good reasons why I have avoided memetic language throughout this book. The first, as I have said, is that it is an effective barrier to communication with those who regard memes as unscientific, speculative and tainted by a view of biological evolution with which they disagree. The idea that the origin of culture can be explained by means of natural and artefactual languages – that the cumulative diversity and complexity of modern human culture can be understood properly only once we begin to ask questions about how humans are able to preserve and exchange sufficient quantities of cultural information, and that the answers to those questions lie in the evolution of our shared systems of discrete representation – offers so much explanatory value that it would be ridiculous to prevent people from taking it seriously simply because I insist on referring to the discrete cultural units as memes.

Another good reason for pushing my memetic proclivities back into the closet is that, though in many ways memetic language enables us to package our concepts in a variety of useful ways, each of those packages is accompanied by some less helpful theoretical baggage. The handy distinction between information and its vehicles, for example, has misled several authors to regard anything as a meme vehicle so long as it travels between humans. Similarly, the claim that a meme is a particulate unit of information "like" a gene draws attention to many aspects of the analogy between biology and culture that I, for one, would not have spotted on my own; but conversely it has some potentially distracting connotations. As in any other use of language, the descriptions that we use when talking about memes are bound to shape the ways in which we think about how culture behaves. It is for this reason, I suspect, that some have been tempted to talk about memes as though they could "leap" from brain to brain, and not to ask important questions about *how* memes might be able to achieve this.

Perhaps the most significant risk that memetic language runs is of giving the impression, when we refer to memes (and indeed genes) as replicators, that these units of information are somehow able, unaided, to create endless copies of themselves. In reality, of course, both genes and memes rely on external mechanisms for their interpretation, expression and replication. Many genes are on a practical level inseparable from these mechanisms, because the genetic information resides in the nucleus of the cells that provide this machinery, whereas memes, like biological viruses, are often physically separate from the machinery on which they rely for their function and reproduction. What matters for evolution, however, is that no information, of any kind. can be inherited in isolation from a receiver that can interpret and implement its prescriptions – and this means that it would perhaps be more accurate to refer to both memes and genes as replicas rather than as replicators (Deacon 1999: 4). This is not to deny Dawkins's insight that genetic information effectively contains "copy me" instructions, but it is to claim that this description is incomplete. Instructions, as we have seen in this book's analysis of heritable information, are always issued to a receiver. Information, like beauty, is in the eye of the beholder, and if there is no beholder then there is, in effect, no information.

So although, in a way, I feel that it is a shame not to make use of memetic shorthand when talking about cultural evolution, I feel more strongly that it is no longer worth the risk of obscuring my words behind the mists of hostility or unwanted implications. Many of this book's arguments could have been redescribed memetically, and there is no representational reason why they should not have been. Nonetheless, there are several sound communicational reasons why it has been better to shed my theory's old memetic skin and focus, instead, on providing sound empirical evidence for the origin of culture by means of natural and artefactual language, and on demonstrating the explanatory value of this thesis by appealing to evidence from a variety of natural and social sciences. For now, at least, even though in my view memetics has established that it is quite theoretically respectable, in practice it is not yet quite socially acceptable.

Acknowledgements

This book is an existence proof of culture's dependence on interactions between many different individuals. Its publication has relied, of course, on the collective expertise of the Cambridge University Press. I am especially grateful to my editor, Beatrice Rehl, for her gentle encouragement and for the clarity of her advice. My thanks are also due to three anonymous readers for their thoughts and suggestions about earlier drafts.

I hold If Price largely responsible for the inception of this book. It was he who started to distract me, with interesting memetic ideas, from the other work that I was doing at the time. The evolution of my thinking about artefactual languages, and about the relationship between language, thought and culture, among other subjects, is the product of many stimulating discussions with If. I have tried to ensure that his contributions have been recognised wherever they have occurred, but his influence on the shape of this book is so pervasive that I am certain to have missed some, for which I apologise.

I am indebted to the many scholars who have been so generous in sharing their publications and thoughts with me. My understanding of the central concepts of cellular biology has been enormously improved by L. S. Shashidhara's responses to the stream of questions that poured into his inbox over the course of several months: Shashi's patience in answering my queries was matched only by the thoroughness and clarity of his replies. Jim Hurford helped me better to understand his own theory of language evolution, and shared with me an unpublished version of his excellent book *The Origins of Meaning*; he is also foremost among the many researchers to whom I am indebted for making their publications available online. John Wilkins helped me to become more familiar with some of the controversies that enmesh the concept of a biological species.

Others who were generous in taking the time to respond to my requests for copies of their publications, clarification of their ideas, comments on my own – or in some cases all three – include Andrew and Rebecca Berkley, Walter Bock, Ted Cloak, Ethan Cochrane, Tom Curtis, Benjamin Dickins, Eric Dietrich, W. Ford Doolittle, Richard Easterlin, Brian Hall, David Hull, Peter Jordan, Amanda Korstjens, James Loy, Arthur Markman, Lynne McClure, Mark Pagel, Karl Magnus Petersson, Joseph Renzulli, Marc van Regenmortel, Gad Yair and Yujian Zheng. My thanks to them all.

This book has benefited greatly from discussions that I have enjoyed with many people, including the late Andor Gomme. The Memetics discussion list hosted a number lively debates, and it was great fun to swap ideas with contributors as diverse as Richard Brodie, Scott Chase, Ted Dace, Ben Dawson, Robin Faichney, Derek Gatherer, Bill Hall, Keith Henson, Alan Patrick, If Price, Ray Recchia, Tim Rhodes, Bill Spight, Chris Taylor and John Wilkins.

Like all of my writing, this book is the fruit of work that has been encouraged by many friends and family members, for which I hope that they all know how grateful I am. It is not enough for a writer to have ideas: she must also have the courage to express them in writing, the motivation to keep rewriting them until they are represented as accurately as possible and the willingness to change them in response to new evidence. All of this work is made immeasurably easier by the support of people who take a genuine interest in it, and in this respect I have been greatly blessed – not least by Calum and Luke, whose enthusiasm about my writing is deeply touching and sustaining, especially given their own preference for books that are rather more action-packed than the ones that their mother produces.

My husband, Keith, has yet again withstood conversations about cultural evolution over meals and evening bottles of wine, on family walks and car journeys; he has tolerated my sprawling stints on the family computer, and the shifting impact of all this on my focus and mood. Those who know him well will detect the footprints of his contributions throughout this book, perhaps in even more places than I have noticed them. Those who know me well will be aware that none of this would have been possible without him.

Finally, there is something to ponder, for those who feel inclined to read the Bible as well as this book, in some verses from the New Testament (italics mine): "At the beginning God *expressed* himself. That *personal expression,* that *word*"[1] – Jesus Christ – "is the radiance of God's glory and the *exact representation* of his being."[2]

[1] John 1:1 (J. B. Phillips, *The New Testament in Modern English,* Inspirational Press, 1972).

[2] Hebrews 1:3 (J. B. Phillips, *The New Testament in Modern English,* Inspirational Press, 1972).

Bibliography

Adler, Nancy, & Stewart, Judith, 2007, *The MacArthur scale of subjective social status*, John D. and Catherine T. MacArthur Research Network on Socioeconomic Status and Health.

Agassi, Joseph, 2006, The biology of the interest in money, *Behavioural and Brain Sciences* 29 (2): 176.

Alexander, Ruth, 2008, The maths of Eurovision voting, *BBC News Magazine*, May 19, http://news.bbc.co.uk/1/hi/7408216.stm.

Andreosso-O'Callaghan, Bernadette, 2002, *Human capital accumulation and economic growth in Asia*, Paper prepared for the Workshop on Asia-Pacific Studies in Australia and Europe: A Research Agenda for the Future, Australian National University, July 5–6, 2002.

Anway, Matthew D., & Skinner, Michael K., 2006, Epigenetic transgenerational actions of endocrine disruptors, *Endocrinology* 147 (6): 43–9.

Argyris, C., 1985, *Strategy, change and defensive routines*, Pitman Publishing.

Aronoff, Stan, & Kaplan, Audrey, 1995, *Total workplace performance: Rethinking the office environment*, WDL Publications.

Atran, Scott, 2001, The trouble with memes: Inference versus imitation in cultural creation, *Human Nature* 12 (4): 351–81.

Aunger, Robert, 2002, *The electric meme: A new theory of how we think*, Free Press.

Aunger, Robert, 2006, An agnostic view of memes, in Jonathan Wells, Simon Strickland & Kevin Laland, eds., *Social information transmission and human biology*, Taylor & Francis, pp. 89–97.

Aunger, Robert, 2007, Memes, in Leslie Barrett and Robin Dunbar, eds., *The Oxford handbook of evolutionary psychology*, Oxford University Press, pp. 599–604.

Aureli, F., Schaffner, C. M., Boesch, C., Bearder, S. K., Call, J., Chapman, C. A., Connor, R., et al., 2008, Fission-fusion dynamics: New research frameworks, *Current Anthropology* 49 (4): 627–54.

Ball, L. Andrew, 2005, Introduction to universal virus taxonomy, in C. M. Fauquet, M. A. Mayo, J. Maniloff, U. Desselberger & L.A. Ball, eds., *Virus taxonomy: VIIIth report of the international committee on taxonomy of viruses*, Academic Press, pp. 3–8.

Barclay, Robert M. R., & Kurta, Allen, 2007, Ecology and behaviour of bats roosting in tree cavities and under bark, in Michael J. Lacki, John P. Hayes & Allen Kurta, eds., *Bats in forests: Conservation and management*, Johns Hopkins University Press, pp. 17–60.

Baron-Cohen, Simon, 2003, *The essential difference*, Penguin Books.

Baron-Cohen, Simon, 2005, The truth about science and sex, *Guardian*, January 27, http://www.guardian.co.uk/science/2005/jan/27/science.educationsgendergap.

Bateson, Gregory, 1972, Form, substance and difference, in *Steps to an ecology of mind*, University of Chicago Press, pp. 448–66.

Benet-Martínez, Verónica, 2006, *Culture and personality research*, http://biculturalism.ucr.edu/research.html.

Bentley, R. Alexander, Hahn, Matthew W., & Shennan, Stephen J., 2004, Random drift and culture change, *Proceedings of the Royal Society of London B – Biological Sciences* 271: 1443–50.

Berk, L., & Garvin, R., 1984, Development of private speech among low-income Appalachian children, *Developmental Psychology* 20 (2): 271–86.

Bhogal, Vicky, 2003, *Cooking like Mummyji: Real British Asian cooking*, Simon & Schuster.

Bilkó, A, Altbäcker, V., & Hudson, R., 1994, Transmission of food preference in the rabbit: The means of information transfer, *Physiology and Behaviour* 56 (5): 907–12.

Bird, Adrian, 2007, Perceptions of epigenetics, *Nature* 447: 396–8.

Blackburn, Simon, 2007, Letter, *London Review of Books* 29 (21), http://www.lrb.co.uk/v29/n21/letters.

Blackburn, Simon, Coyne, Jerry, Kitcher, Philip, Lewens, Tim, & Rose, Steven, 2008, Letter, *London Review of Books* 30 (1). http://www.lrb.co.uk/v30/n01/letters.html.

Blackmore, Susan, 1999, *The meme machine*, Oxford University Press.

Blackmore, Susan, 2006, Memetics by another name? *Bioscience* 56: 74–5.

Blashfield, Roger K., & Flanagan, Elizabeth H., 2005, A language and methodology for studying the hierarchies in the DSMs, in Stephen Strack & Theodore Millon, eds., *Handbook of personology and psychopathology*, John Wiley & Sons, pp. 50–72.

Bloch, Maurice, 2000, A well-disposed social anthropologist's problems with memes, in Robert Aunger, ed., *Darwinizing culture: The status of memetics as a science*, Oxford University Press, pp. 189–204.

Blute, Marion, 2002, The evolutionary ecology of science, *Journal of Memetics – Evolutionary Models of Information Transmission* 7 (1), http://cfpm.org/jom-emit/2003/vol7/blute_m.html.

Blute, Marion, 2005, Memetics and evolutionary social science, *Journal of Memetics – Evolutionary Models of Information Transmission* 9 (1), http://jom-emit.cfpm.org/2005/vol9/blute_m.html.

Blute, Marion, 2007, *The role of memes in cultural evolution: Memes if necessary, but not necessarily memes*, http://www.semioticon.com/virtuals/imitation/mblute_paper.pdf.

Bock, Walter J., 1995, The species concept versus the species taxon: Their roles in biodiversity analyses and conservation, in R. Arai, M. Kato & Y. Doi, eds., *Biodiversity and evolution*, National Science Museum Foundation, Tokyo, pp. 47–72.

Bock, Walter J., 2004a, Species: The concept, category and taxon, *Journal of Zoological Systematics and Evolutionary Research* 42 (3): 178–90.

Bock, Walter J., 2004b, Explanations in systematics, in David M. Williams & Peter L. Forey, eds., *Milestones in systematics: The systematics association special volume series 67*, CRC Press, pp. 49–56.

Bogoshi, Jonas, Naidoo, Kevin, & Webb, John, 1987, The oldest mathematical artefact, *Mathematical Gazette* 71 (458): 294.

Boissac, Paul, 2006, Hoarding behavior: A better evolutionary account of money psychology? *Behavioural and Brain Sciences* 29: 181–2.

Booth, Alan, 1979, Does wives' employment cause stress for husbands? *Family Coordinator* 28 (4): 445–9.

Bortfeld, Heather, 2002, *Neural correlates of infant word recognition*, James S. McDonnell Foundation.

Botterill, George, 1993, *Functions and functional explanation*, Unpublished manuscript, Department of Philosophy, University of Sheffield.

Bourhis, R. Y., & Giles, H., 1977, The language of intergroup distinctiveness, in H. Giles, ed., *Language, ethnicity and intergroup relations*, Academic Press, pp. 119–34.

Boyd, R., & Richerson, Peter J., 1985, *Culture and the evolutionary process*, University of Chicago Press.

Boyd, R., & Richerson, Peter J., 2000, Meme theory oversimplifies cultural change, *Scientific American*, October, 54–5.

Boyd, R., & Richerson, Peter J., 2006, Culture and the evolution of the human social instincts, in S. Levinson & N. Enfield, eds., *Roots of human sociality*, Berg Publishers, pp. 453–77.

Brandon, Robert N., 1996, *Concepts and methods in evolutionary biology*, Cambridge University Press.

Briers B., Pandelaere M., Dewitte S., & Warlop L., 2006, Hungry for money: The desire for caloric resources increases the desire for financial resources and vice versa? *Psychological Science* 17: 939–43.

Brigandt, Ingo, 2003, Species pluralism does not imply species eliminativism, *Philosophy of Science* 70: 1305–16.

Brighton, Henry, & Kirby, Simon, 2001, The survival of the smallest: Stability conditions for the cultural evolution of compositional language, in J. Kelemen & P. Sosík, eds., *Proceedings of the European Conference on Artificial Life (ECAL01)*, Prague, Springer-Verlag, pp. 592–601.

Brockwood, Krista J., 2007, Marital satisfaction and the work-family interface: An overview, in Stephen Sweet & Judi Casey, eds., *Work and family encyclopedia*, http://wfnetwork.bc.edu/encyclopedia_entry.php?id=4236&area=All.

Bruner, J. S., & Goodman, C. C., 1947, Value and needs as organizing factors in perception, *Journal of Abnormal and Social Psychology* 42: 33–44.

Byrne, R., & Whiten, A., eds., 1988, *Machiavellian intelligence*, Oxford University Press.

Byrne, Richard W., & Russon, Anne E., 1998, Learning by imitation: A hierarchical approach, *Behavioural and Brain Sciences* 21 (5): 667–8.

Calhoun, Ada, 1999, Count her in: Denise Schmandt-Besserat's new way of seeing, *Austin Chronicle*, December 13, http://www.weeklywire.com/ww/12-13-99/austin_books_feature.html.

Call, Josep, Carpenter, Malinda, & Tomasello, Michael, 2005, Copying results and copying actions in the process of social learning: Chimpanzees (*Pan troglodytes*) and human children (*Homo sapiens*), *Animal Cognition* 8: 151–63.

Carruthers, Peter, 1996, Autism as mind-blindness: An elaboration and partial defence, in P. Carruthers & P. K. Smith, eds., *Theories of theories of mind*, Cambridge University Press, pp. 257–73.

Cavalli, Giacomo, & Paro, Renato, 1998, The Drosophila Fab-7 chromosomal element conveys epigenetic inheritance during mitosis and meiosis, *Cell* 93 (4): 505–18.

Chandler, Daniel, 1994, *The transmission model of communication*, http://www.aber.ac.uk/media/Documents/short/trans.html.

Chaney, Richard Paul, 1978, Polythematic expansion: Remarks on Needham's polythetic classification, *Current Anthropology* 19 (1): 139–43.

Cheney, D. L., & Seyfarth, R., 1990, *How monkeys see the world: Inside the mind of another species*, University of Chicago Press.

Chomsky, Noam, 1987, Language: Chomsky's theory, in Richard L. Gregory, ed., *The Oxford companion to the mind*, Oxford University Press, pp. 419–21.

Christiansen, Morten H., & Kirby, Simon, eds., 2003a, *Language evolution*, Oxford University Press.

Christiansen, Morten H., & Kirby, Simon, 2003b, Language evolution: Consensus and controversies, *Trends in Cognitive Sciences* 7 (7): 300–7.

Clark, Andy, 1998a, Magic words? How language augments human computation, in P. Carruthers & J. Boucher, eds., *Language and thought: Interdisciplinary themes*, Cambridge University Press, pp. 162–83.

Clark, Andy, 1998b, Twisted tales: Causal complexity and cognitive scientific explanation, *Minds and Machines* 8: 79–99. (Reprinted in F. Keil & R. A. Wilson, eds., 2000, *Explanation and Cognition*, MIT Press, pp. 145–66.)

Clark, Andy, & Chalmers, David J., 1998, The extended mind, *Analysis* 58 (1): 7–19.

Clay, Rebecca A., 2001, Wealth secures health, *Monitor on Psychology* 32 (9): 78.

Clerides, Sofronis, & Stengos, Thanasis, 2006, Love thy neighbor, love thy kin: Strategy and bias in the Eurovision Song Contest, *Social Science Research Network*, May 2006, http://ssrn.com/abstract=882383.

Cloak, F. T., 1966, Cultural microevolution, *Research Previews* 13: 7–10.

Cloak, F. T., 1968, *Cultural Darwinism: Natural selection of the spoked wood wheel*, Transcript of an automated slide presentation "The Wheel" at the Annual Meeting of the American Anthropological Association, Seattle, November 1968.

Cloak, F. T., 1975a, Is a cultural ethology possible? *Human Ecology* 3 (3): 161–82.

Cloak, F. T., 1975b, That a culture and a social organization mutually shape each other through a process of continuing evolution, *Man-Environment Systems* 5: 3–6.

Cloak, F. T., 2008, *Perceptual control theory (PCT) and the on-going evolution of culture*, Unpublished manuscript, http://www.tedcloak.com/.

Cochrane, Ethan E., 2007, *How useful is memetics to evolutionary archeologies?* Paper presented at the symposium "Imitation, memory, and cultural changes: Probing the meme hypothesis," Victoria College, Toronto, May 4–6, 2007.

Cohen, Sheldon, 1999, Social status and susceptibility to respiratory infections, *Annals of the New York Academic of Sciences* 896: 246–53.

Connor, R. C., Wells, R., Mann, J., & Read, A., 2000, The bottlenose dolphin: Social relationships in a fission-fusion society, in J. Mann, R. C. Connor, P. Tyack & H. Whitehead, eds., *Cetacean societies: Field studies of dolphins and whales*, University of Chicago Press, pp. 91–126.

Conte, Rosaria, & Paolucci, Mario, 2001, Intelligent social learning, *Journal of Artificial Societies and Social Stimulation* 4 (1), http://www.soc.surrey.ac.uk/JASSS/4/1/3.html.

Cowan, N., 2001, The magical number 4 in short-term memory: A reconsideration of mental storage capacity, *Behavioural and Brain Sciences* 24: 87–114.

Coyne, Jerry, & Kitcher, Philip, 2007, Letter, *London Review of Books* 29 (22), http://www.lrb.co.uk/v29/n22/letters.

Cross, Ian, 2007, Letter, *London Review of Books* 29 (21), http://www.lrb.co.uk/v29/n21/letters.html.

Crosson, B., Moberg, P. J., Boone, J. R., Rothi, L. J. G., & Raymer, A., 1997, Category-specific naming deficit for medical terms after dominant thalamic/capsular hemorrhage, *Brain and Language* 60: 407–42.

Curtis, Thomas P., Head, Ian M., Lunn, Mary, Schloss, Patrick D., & Sloan, William T., 2006, What is the extent of prokaryotic diversity? *Philosophical Transactions of the Royal Society B* 361: 2023–37.

Curtis, Thomas P., & Sloan, William T., 2004, Exploring microbial diversity: A vast below, *Science* 309: 1331–3.

Curtis, Thomas P., Wallbridge, Nigel C., & Sloan, William T., 2008, Theory, community assembly, diversity and evolution in the microbial world, in Roger K. Butlin, Jon R. Bridle & Dolph Schluter, eds., *Speciation and patterns of diversity*, Cambridge University Press, pp. 59–76.

Damerow, Peter, 1999, The origins of writing as a problem of historial epistemology, invited lecture at Center for Ancient Studies Symposium, "The Origins of Writing: Image, Symbol and Script," University of Pennsylvania, March 26–27.

Darwin, Charles, 1859, *The origin of species by means of natural selection*, Penguin Books, 1985.

Darwin, Charles, 1871, *The descent of man, and selection in relation to sex*, Princeton University Press, 1981.

Davies, Glyn, 1997, *Monetary innovation in historical perspective*, keynote address at the first Consult Hyperion Digital Money Forum, London, October 7–8.

Davies, Glyn, 2002, *A history of money from ancient times to the present day*, University of Wales Press.

Davies, Ron, 2005, *Origins of money and of banking*, http://www.ex.ac.uk/~RDavies/arian/origins.html.

Dawkins, Richard, 1976, *The selfish gene*, Oxford University Press.

Dawkins, Richard, 1982, *The extended phenotype: The long reach of the gene*, Oxford University Press.

Dawkins, Richard, 1984, Replicators and vehicles, in Robert N. Brandon & Richard M. Burian, eds., *Genes, organisms, populations: Controversies over the units of selection*, Massachusetts Institute of Technology Press, pp. 161–80.

Dawkins, Richard, 1986, *The blind watchmaker*, Penguin Books.

Dawkins, Richard, 1996, *Climbing mount improbable*, Viking.

Dawkins, Richard, 2003, *A devil's chaplain: Selected essays*, Phoenix.

Dawkins, Richard, 2004a, Extended phenotype – but not too extended: A reply to Laland, Turner and Jablonka, *Biology and Philosophy* 19: 377–96.

Dawkins, Richard, 2004b, Gerin oil, *Free Inquiry* 24 (1): 9–11.

De Angelis, Gessica, & Selinker, Larry, 2001, Interlanguage transfer and competing linguistic systems in the multilingual mind, in Jasone Cenoz, Britta Hufeisen & Ulrike Jessner, eds., *Third language acquisition: Psycholinguistic perspectives*, Multilingual Matters, pp. 42–58.

de la Cruz, Fernando, & Davies, Julian, 2000, Horizontal gene transfer and the origin of species: Lessons from bacteria, *Trends in Microbiology* 8: 128–33.

Deacon, Terrence W., 1992, Brain-language coevolution, in John A. Hawkins & Murray Gell-Mann, eds., *The evolution of human languages*, Addison-Wesley, pp. 49–83.

Deacon, Terrence W., 1997, *The symbolic species: The co-evolution of language and the human brain*, Allen Lane, Penguin Press.

Deacon, Terrence W., 1999, Editorial: Memes as signs, *Semiotic Review of Books* 10 (3): 1–3.

Dennett, Daniel, 1991, *Consciousness explained*, Penguin Books.

Dennett, Daniel, 1995, *Darwin's dangerous idea: Evolution and the meanings of life*, Allen Lane, Penguin Press.

Dennett, Daniel, 1998, *Memes: Myths, misunderstandings and misgivings*, paper presented at the Chapel Hill Colloquium, North Carolina, October 1998.

Dennett, Daniel, 2006a, *Breaking the spell: Religion as a natural phenomenon*, Allen Lane.

Dennett, Daniel, 2006b, From typo to thinko: When evolution graduated to semantic norms, in Stephen C. Levinson & Pierre Jaisson, eds., *Evolution and culture*, Massachusetts Institute of Technology Press, pp. 133–45.

Dennett, Daniel, 2007, Letter, *London Review of Books* 29 (22), http://www.lrb.co.uk/v29/n22/letters.

Dennett, Daniel, & McKay, Ryan, 2006, A continuum of mindfulness, *Behavioral and Brain Sciences* 29: 353–4.

Devlin, Keith, 1999, *Infosense: Turning information into knowledge*, W. H. Freeman.

Dickins, Thomas E., 2005, On the aims of evolutionary theory, *Evolutionary Psychology* 3: 79–84.

Dickins, Thomas E., & Dickins, Benjamin J. A., 2008, Mother nature's tolerant ways: Why non-genetic inheritance has nothing to do with evolution, *New Ideas in Psychology* 26: 41–54.

Dietrich, Eric, 2000, Analogy and conceptual change, or, you can't step into the same mind twice, in E. Dietrich & A. Markman, eds., *Cognitive dynamics: Conceptual change in humans and machines*, Lawrence Erlbaum, pp. 265–94.

Dietrich, Eric, & Markman, Arthur B., 2003, Discrete thoughts: Why cognition must use discrete representations, *Mind and Language* 18 (1): 95–119.

Dirlam, David K., 2005, Using memetics to grow memetics, *Journal of Memetics – Evolutionary Models of Information Transmission* 9 (1), http://cfpm.org/jom-emit/2005/vol9/dirlam_dk.html.

Distin, K. E., 1997, Cultural evolution: The meme hypothesis, in *Proceedings of the Conference on Evolution, or How Did We Get Here?* Westminster College Oxford, Christ and the Cosmos.

Distin, Kate, 2005, *The selfish meme: A critical reassessment*, Cambridge University Press.

Distin, Kate, 2006a, Fame! Money! Power! *Think: Philosophy for Everyone* 13: 89–93.

Distin, Kate, ed., 2006b, *Gifted children: A guide for parents and professionals*, Jessica Kingsley Publishers.

Distin, K. E., & Distin, K. W., 1996, Human design methods, in *Proceedings of the Conference on Design in Nature?* Birmingham, October 5, 1996.

Dodd, Nigel, 2004, *The nature of money*, Polity Press.

Dodd, Nigel, 2005, Reinventing monies in Europe, *Economy and Society* 34 (4): 558–83.

Doolittle, W. Ford, 1999, Phylogenetic classification and the universal tree, *Science* 284: 2124–8.

Doolittle, W. Ford, & Bapteste, Eric, 2007, Pattern pluralism and the tree of life hypothesis, *Proceedings of the National Academy of Sciences* 104 (7): 2043–9.

Dunbar, Robin I. M., 1998, The social brain hypothesis, *Evolutionary Anthropology* 6:178–90.

Dunbar, Robin I. M., 1999, Culture, honesty and the freerider problem, in R. Dunbar, C. Knight & C. Power, eds., *The evolution of culture*, Edinburgh University Press, pp. 194–213.

Dunbar, Robin I. M., 2000, On the origin of the human mind, in Peter Carruthers & Andrew Chamberlain, eds., *Evolution and the human mind: Modularity, language and meta-cognition*, Cambridge University Press, pp. 238–53.

Dunbar, Robin I. M., 2003, The social brain: Mind, language and society in evolutionary perspective, *Annual Review of Anthropology* 32: 163–81.

Durkheim, Émile, 1938, *The rules of sociological method*, University of Chicago Press.

Durkheim, Émile, 1964, *The division of labor in society*, Free Press.

Duthie, A. Bradley, 2004, The fork and the paperclip: A memetic perspective, *Journal of Memetics – Evolutionary Models of Information Transmission* 8 (1), http://cfpm.org/jom-emit/2004/vol8/duthie_ab.html.

Easterlin, R. A., 1981, Why isn't the whole world developed? *Journal of Economic History* 41 (1): 1–19.

Edmonds, Bruce, 2005, The revealed poverty of the gene-meme analogy: Why memetics per se has failed to produce substantive results, *Journal of Memetics – Evolutionary Models of Information Transmission* 9 (1), http://cfpm.org/jom-emit/2005/vol9/edmonds_b.html.

Eerkens, Jelmer W., & Lipo, Carl P., 2007, Cultural transmission theory and the archaeological record: Providing context to understanding variation and temporal changes in material culture, *Journal of Archaeological Research* 15: 239–74.

Emery, Nathan J., Lorincz, Erika N., Perrett, David I., Oram, Michael W., & Baker, Christopher I., 1997, Gaze following and joint attention in Rhesus monkeys (*Macaca mulatto*), *Journal of Comparative Psychology* 111 (3): 286–93.

Enfield, Nicholas J., & Levinson, Stephen C., eds., 2006, Introduction: Human sociality as a new interdisciplinary field, in Nicholas J. Enfield & Stephen C. Levinson, eds., *Roots of human sociality: Culture, cognition and interaction*, Berg Publishers.

Epel, Elissa, Lapidus, Rachel, McEwen, Bruce, & Brownell, Kelly, 2001, Stress may add bite to appetite in women: A laboratory study of stress-induced cortisol and eating behavior, *Psychoneuroendocrinology* 26 (1): 37–49.

Ereshefsky, Marc, 2002, *Species*, Stanford Encyclopedia of Philosophy, http://plato.stanford.edu/entries/species.

Evans, Jonathan St. B. T., 2003, In two minds: Dual-process accounts of reasoning, *Trends in Cognitive Sciences* 7 (10), pp. 454–9.

Fedak, T., Franz-Odendaal, T., Hall, B. K., & Vickaryous, M., 2004, Epigenetics: The context of development, *Paleontological Association Newsletter* 56: 44–9.

Fodor, Jerry, 2007a, Letter, *London Review of Books* 29 (23), http://www.lrb.co.uk/v29/n23/letters.html.

Fodor, Jerry, 2007b, Why pigs don't have wings, *London Review of Books* 29 (20): 19–22.

Frankel, Jeffrey A., & Rose, Andrew K., 2002, Estimating the effect of currency unions on trade and output, *Quarterly Journal of Economics* 117 (2): 437–66.

Furnham, Adrian, 1983, Inflation and the estimated sizes of notes, *Journal of Economic Psychology* 4(4): 349–52.

Gabora, L., & Aerts, D., 2005, Distilling the essence of an evolutionary process and implications for a formal description of culture, in W. Kistler, ed., *Proceedings of the Center for Human Evolution Workshop #5: Cultural Evolution*, May 2000, Foundation for the Future.

Gabora, Liane, 1996, A day in the life of a meme, *Philosophica* 57: 901–38

Gabora, Liane, 2006, The fate of evolutionary archaeology: Survival or extinction? *World Archaeology* 38 (4): 690–6.

Galef, Bennett G., 2004, Approaches to the study of traditional behaviors of free-living animals, *Learning and Behavior* 32 (1): 53–61.

Gatherer, Derek, 1997, Macromemetics: Towards a framework for the re-unification of philosophy, *Journal of Memetics – Evolutionary Models of Information Transmission* 1 (1), http://cfpm.org/jom-emit/1997/vol1/gatherer_dg.html.

Gatherer, Derek, 2006a, Comparison of Eurovision Song Contest simulation with actual results reveals shifting patterns of collusive voting alliances, *Journal of Artificial Societies and Social Simulation* 9 (2), http://jasss.soc.surrey.ac.uk/9/2/1/1.pdf.

Gatherer, Derek, 2006b, Cultural evolution: The biological perspective, *Parallax* 12 (1): 57–68.

Gergely, György, Bekkering, Harold, & Király, Ildikó, 2002, Rational imitation in preverbal infants, *Nature* 415: 755.

Gergely, György, & Csibra, Gergely, 2005, The social construction of the cultural mind: Imitative learning as a mechanism of human pedagogy, *Interaction Studies* 6 (3): 463–81.

Gibbs, Christopher H., 2006, Notes on Beethoven's fifth symphony, *Philadelphia Orchestra Association programme note*, http://www.npr.org/templates/story/story.php?storyId=5473894.

Giles, H., & Powesland, P. F., 1975, *Speech style and social evaluation (European monographs in social psychology 7)*, Academic Press.

Giles, H., & Smith, P., 1979, Accommodation theory: Optimal levels of convergence, in H. Giles & R. N. St Clair, eds., *Language and social psychology*, Basil Blackwell, pp. 45–65.

Ginsburgh, Victor, & Noury, Abdul, 2008, The Eurovision Song Contest: Is voting political or cultural? *European Journal of Political Economy* 24 (1): 41–52.

Girouard, Mark, 1980, *Life in the English country house: A social and architectural history*, Penguin Books.

Glick, Reuven, & Rose, Andrew K., 2002, Does a currency union affect trade? The time series evidence, *European Economic Review* 46 (6): 1125–51.

Godfrey-Smith, Peter, & Sterelny, Kim, 2007, Biological information, in Edward N. Zalta, ed., *Stanford encyclopedia of philosophy*, http://plato.stanford.edu/archives/win2007/entries/information-biological/.

Goodwin, Charles, 2006, Human sociality as mutual orientation in a rich interactive environment: Multimodal utterances and pointing in aphasia, in Nicholas J. Enfield & Stephen C. Levinson, eds., *Roots of human sociality: Culture, cognition and interaction*, Berg Publishers, pp. 97–125.

Gray, Russell D., Greenhill, Simon J., & Ross, Robert M., 2008, The pleasures and perils of Darwinizing culture (with phylogenies), *Biological Theory* 2 (4): 360–75.

Griffiths, Paul E., 2000, David Hull's natural philosophy of science, *Biology and Philosophy* 15: 301–10.

Griffiths, Paul E., 2003, Beyond the Baldwin effect: James Mark Baldwin's "social heredity," epigenetic inheritance and niche-construction, in Bruce Weber & David Depew, eds., *Evolution and learning: The Baldwin effect reconsidered*, Massachusetts Institute of Technology Press, pp. 193–215.

Griffiths, Paul E., & Gray, Russell D., 2004, The developmental systems perspective: Organism environment systems as units of development and evolution, in K. Preston & M. Pigliucci, eds., *Phenotypic integration: Studying the ecology and evolution of complex phenotypes*, Oxford University Press, pp. 409–431.

Hall, B. K., 1998, Epigenetics: Regulation not replication, *Journal of Evolutionary Biology* 11: 201–5.

Hall, William P., 2006, *Tools extending human and organizational cognition: Revolutionary tools and cognitive revolutions*, 6th International Conference on Knowledge, Culture and Change in Organisations, Prato, Italy, July 11–14.

Hall, William P., 2007, *Application holy wars or a new reformation? A fugue on the theory of knowledge*, http://www.orgs-evolution-knowledge.net/Index/ApplicHolyWars/FullProjectToNow/ApplicationHolyWarsWeb/default.htm.

Henrich, Joseph, & Boyd, Robert, 2002, On modeling cognition and culture: Why cultural evolution does not require replication of representations, *Culture and Cognition* 2: 87–112.

Henrich, Joseph, Boyd, R., & Richerson, Peter J., 2008, Five misunderstandings about cultural evolution, *Human Nature* 19 (2): 119–37.

Higgins, E. T., 1996, Knowledge activation: Accessibility, applicability and salience, in E. T. Higgins & A. E. Kruglanski, eds., *Social psychology: Handbook of basic principles*, Guilford Press, pp. 133–68.

Hitchins, M. P., Wong, J. J., Suthers, G., Suter, C. M., Martin, D. I., Hawkins, N. J., & Ward, R. L., 2007, Inheritance of a cancer-associated MLH1 germ-line epimutation, *New England Journal of Medicine* 356 (7): 697–705.

Honderich, Ted, ed., 1995, *The Oxford companion to philosophy*, Oxford University Press.

Hong, Ying-Yi, Morris, Michael W., Chiu, Chi-yue, & Benet-Martínez, Véronica, 2000, Multicultural minds: A dynamic constructivist approach to culture and cognition, *American Psychologist* 55 (7): 709–20.

Howe, Carl, 2006, The *Wall St. Journal*'s faulty conclusion from the VHS-Betamax war (SNE), *Seeking Alpha*, January 26, 2006, http://seekingalpha.com/article/6178-the-wall-st-journal-s-faulty-conclusion-from-the-vhs-betamax-war-sne.

Howick, Jeremy, 2005, Book review: The selfish meme, *Philosophy Today* 19 (50).

Hull, David, & Wilkins, John S., 2005, Replication, in Edward N. Zalta, ed., *Stanford encyclopedia of philosophy*, http://plato.stanford.edu/archives/spr2005/entries/replication/.

Hull, David L., 1978, Altruism in science: A sociobiological model of co-operative behaviour among scientists, *Animal Behaviour* 26: 685–97.

Hull, David L., 1982, The naked meme, in H. C. Plotkin, ed., *Learning, development and culture*, John Wiley & Sons, pp. 273–327.

Hull, David L., 1988, Interactors versus vehicles, in Henry C. Plotkin, ed., *The role of behavior in evolution*, Massachusetts Institute of Technology Press, pp. 19–50.

Humle, Tatyana, & Matsuzawa, Tetsuro, 2002, Ant-dipping among the chimpanzees of Bossou, Guinea, and some comparisons with other sites, *American Journal of Primatology* 58 (3): 133–48.

Humphrey, Nicholas, 1998, Cave art, autism, and the evolution of the human mind, *Cambridge Archaeological Journal* 8 (2): 165–91.

Hurford, James R., 1994, Linguistics and evolution: A background briefing for non-linguists, *Current Opinion in Neurobiology* 2: 180–5.

Hurford, James R., 1998, Review of *The symbolic species: The co-evolution of language and the human brain*, by Terrence Deacon, Times Literary Supplement, October 23: 34

Hurford, James R., 1999, The evolution of language and languages, in Robin Dunbar, Chris Knight & Camilla Power, eds., *The evolution of culture*, Edinburgh University Press, pp. 173–93.

Hurford, James R., 2000, The emergence of syntax, in C. Knight, M. Studdert-Kennedy & J. Hurford, eds., *The evolutionary emergence of language: Social function and the origins of linguistic form*, Cambridge University Press, pp. 219–30.

Hurford, James R., 2001, Numeral systems, in N. J. Smelser & P. B. Baltes, eds., *International encyclopedia of the social and behavioral sciences*, Pergamon, pp. 10756–61.

Hurford, James R., 2002, The roles of expression and representation in language evolution, in Alison Wray, ed., *The transition to language*, Oxford University Press, pp. 311–34.

Hurford, James R., 2003a, The language mosaic and its evolution, in Morten Christiansen & Simon Kirby, eds., *Language evolution*, Oxford University Press, pp. 38–57.

Hurford, James R., 2003b, Why synonymy is rare: Fitness is in the speaker, in W. Banzhaf, T. Christaller, P. Dittrich, J. T. Kim & J. Ziegler, eds., *Advances in artificial life: Proceedings of the 7th European Conference on Artificial Life (ECAL), Lecture notes in artificial intelligence*, Springer Verlag, pp. 442–51.

Hurford, James R., 2004, Human uniqueness, learned symbols and recursive thought, *European Review* 12 (4): 551–65.

Hurford, James R., 2007, *The origins of meaning: Language in the light of evolution*, Oxford University Press.

Hurford, James R., & Kirby, Simon, 1999, Co-evolution of language-size and the critical period, in David Birdsong, ed., *Second language acquisition and the critical period hypothesis*, Lawrence Erlbaum, pp. 39–63.

Ingham, Geoffrey, 1996, Money is a social relation, *Review of Social Economy* 54 (4): 507–29.

Ingham, Geoffrey, 2004, *The nature of money*, Polity Press.

Ingham, Geoffrey, 2006, Further reflections on the ontology of money: Responses to Lapavitsas and Dodd, *Economy and Society* 35 (2): 259–78.

Ingham, Geoffrey, 2008, *Capitalism*, Polity Press.

Irwin, Rebecca E., 2001, *The University of Tennessee at Martin Department of Biological Sciences, Zoology 441: Animal Ecology Lecture on Life History Theory*, unpublished lecture notes, http://www.utm.edu/departments/cens/biology/rirwin/441_442/441LifeHistoryTheory.htm.

Jablonka, Eva, & Lamb, Marion J., 2005, *Evolution in four dimensions*, Massachusetts Institute of Technology Press.

Jablonka, Eva, & Lamb, Marion J., 2006, Precis of evolution in four dimensions, *Behaviour and Brain Sciences*, http://www.bbsonline.org/Preprints/Jablonka-10132006/Referees/Jablonka-10132006_Precis.preprint.pdf.

James, P., 2004, Obesity: The worldwide epidemic, *Clinics in Dermatology* 22 (4): 276–80.

Jan, Steven, 2007, *The memetics of music: A neo-Darwinian view of musical structure and culture*, Ashgate Publishing.

Jarrold, Chris, Carruthers, Peter, Smith, Peter K., & Boucher, Jill, 1994, Pretend play: Is it metarepresentational? *Mind and Language* 9 (4): 445–68.

Jasper, H., & Penfield, W., 1954, *Epilepsy and the functional anatomy of the human brain*, 2nd ed., Little, Brown.

Jessner, Ulrike, 1999, Metalinguistic awareness in multilinguals: Cognitive aspects of third language acquisition, *Language Awareness* 8: 201–9.

Johnson, Boris, 2007, Madeleine McCann saga reflects our society, *Daily Telegraph*, September 13, http://www.telegraph.co.uk/comment/3642638/Madeleine-McCann-saga-reflects-our-society.html.

Jordan, Peter, & Shennan, Stephen, 2002, Cultural transmission, language, and basketry traditions amongst the California Indians, *Journal of Anthropological Archaeology* 22: 42–74.

Jorion, Paul, 2006, Keeping up with the Joneses: The desire of the desire for money, *Behavioural and Brain Sciences* 29: 187–8.

Kaminski, Juliane, Call, Josep, & Fischer, Julia, 2004, Word learning in a domestic dog: Evidence for "fast mapping," *Science* 304: 1682–3.

Kasser, T., & Ryan, R. M., 1993, A dark side of the American dream: Correlates of financial success as a central life aspiration, *Journal of Personality and Social Psychology* 65: 410–22.

Khosla, Vipul, 2008, *Memes, signs and drama: Case study analysis of malaria communication campaign in Cambodia*, unpublished M.Sc. diss., Department of Media and Communication, London School of Economics.

Kinzler, Katherine D., Dupoux, Emmanuel, & Spelke, Elizabeth S., 2007, The native language of social cognition, *Proceedings of the National Academy of Sciences* 104 (30): 12577–80.

Kirby, Simon, 2007, The evolution of language, in R. Dunbar & L. Barrett, L., eds., *Oxford handbook of evolutionary psychology*, Oxford University Press, pp. 669–81.

Kniffin, Kevin M., 2006, Show me the status: Money as a kind of currency, *Behavioural and Brain Sciences* 29: 188–9.

Koestler, Arthur, 1978, *Janus: A summing up*, Hutchinson.

Kuhl, P. K., Williams, K. A., Lacerda, F., Stevens, K. N., & Lindblom, B., 1992, Linguistic experience alters phonetic perception in infants by 6 months of age, *Science* 255: 606–8.

Kuhl, Patricia K., & Meltzoff, Andrew N., 1997, Evolution, nativism and learning in the development of language and speech, in M. Gopnik, ed., *The inheritance and innateness of grammars*, Oxford University Press, pp. 7–44.

Kuhn, Deanna, 1991, *The skills of argument*, Cambridge University Press.

Labov, W., 1963, The social motivation of a sound change, *Word* 19: 273–309.

Lakhotia, Subhash C., 1997a, What is a gene? 1. A question with variable answers, *Resonance* (April): 38–47.

Lakhotia, Subhash C., 1997b, What is a gene? 2. A question with variable answers, *Resonance* (May): 44–53.

Laland, K. N., Odling-Smee, F. J., & Feldman, M. W., 1999, Evolutionary consequences of niche construction and their implications for ecology, *Proceedings of the National Academy of Sciences* 96 (18): 10242–7.

Lapavitsas, Costas, 2003, *Money as "universal equivalent" and its origin in commodity exchange*, http://mercury.soas.ac.uk/economics/workpap/adobe/wp130.pdf.

Lapavitsas, Costas, 2005a, *The emergence of money in commodity exchange; or, money as monopolist of the ability to buy*, School of Oriental and African Studies Working Paper 126, http://www.soas.ac.uk/economics/research/workingpapers/28852.pdf.

Lapavitsas, Costas, 2005b, The social relations of money as universal equivalent: A response to Ingham, *Economy and Society* 34 (3): 389–403.

Lea, S. E. G., 1981, Inflation, decimalization and the estimated size of coins, *Journal of Economic Psychology* 1: 79–81.

Lea, Stephen E. G., & Webley, Paul, 2006, Money as tool, money as drug: The biological psychology of a strong incentive, *Behavioural and Brain Sciences* 29: 161–76.

Leiser, D., & Izak, G., 1987, The money size illusion as a barometer of confidence? The case of high inflation in Israel, *Journal of Economic Psychology* 8 (3): 347–56.

LePage, R. B., 1968, Problems of description in multilingual communities, *Transactions of the Philological Society* 67 (1): 189–212.

Levinson, Stephen C., 2003, Language and mind: Let's get the issues straight! in D. Gentner & S. Goldin-Meadow, eds., *Language in mind: Advances in the study of language and cognition*, Massachusetts Institute of Technology Press, pp. 25–46.

Levinson, Stephen C., 2006, Introduction: The evolution of culture in a microcosm, in Stephen C. Levinson & Pierre Jaisson, eds., *Evolution and culture: A Fyssen Foundation symposium*, Massachusetts Institute of Technology Press, pp. 1–41.

Lewens, Tim, 2007, Letter, *London Review of Books* 29 (21), http://www.lrb.co.uk/v29/n21/letters.html.

Lewis-Williams, J. D., 2003, Chauvet: The cave that changed expectations, *South African Journal of Science* 99: 191–4.

Li, Norman P., Bailey, J. Michael, Kenrick, Douglas T., & Linsenmeier, Joan A. W., 2002, The necessities and luxuries of mate preferences: Testing the tradeoffs, *Journal of Personality and Social Psychology* 82 (6): 947–55.

Lieberman, M. D., Eisenberger, N. I., Crockett, M. J., Tom, S. M., Pfeifer, J. H., & Way, B. M., 2007, Putting feelings into words: Affect labeling disrupts amygdala activity to affective stimuli, *Psychological Science* 18: 421–8.

Linder, C. Randal, Moret, Bernard M. E., Nakhleh, Luay, & Warnow, Tandy, 2003, *Network (reticulate) evolution: Biology, models and algorithms*, a tutorial presented at the Ninth Pacific Symposium on Biocomputing, January 6–10, 2004, Hawaii, http://www.cs.rice.edu/~nakhleh/Papers/psb04.pdf.

Lipo, Carl P., O'Brien, Michael J., Collard, Mark, & Shennan, Stephen J., 2005, Cultural phylogenies and explanation: Why historical methods matter, in *Mapping our ancestors: Phylogenetic approaches in anthropology and prehistory*, Aldine-Transaction, pp. 3–16.

Liszkowski, Ulf, 2006, Infant pointing at 12 months: Communicative goals, motives, and social-cognitive abilities, in Nicholas J. Enfield & Stephen C. Levinson, eds., *Roots of human sociality: Culture, cognition and interaction*, Berg Publishers, pp. 153–77.

Livesley, W. J., 1985, The classification of personality disorder: 1. The choice of category concept, *Canadian Journal of Psychiatry* 30: 353–8.

Lord, Andrew, & Price, If, 2001, Reconstruction of organisational phylogeny from memetic similarity analysis: Proof of feasibility, *Journal of Memetics – Evolutionary Models of Information Transmission* 5 (2), http://cfpm.org/jom-emit/2001/vol5/lord_a&price_i.html.

Loy, James, 1970, Behavioral responses of free-ranging rhesus monkeys to food shortage, *American Journal of Physical Anthropology* 33 (2): 263–71.

Lukes, Steven, 1975, *Émile Durkheim – His life and work*, Peregrine.

Lyman, R. Lee, & O'Brien, Michael J., 2003, Cultural traits: Units of analysis in early twentieth-century anthropology, *Journal of Anthropological Research* 59: 225–50.

Lynch, Aaron, 1996, *Thought contagion: How belief spreads through society*, Basic Books.

Madden, Joah R., Lowe, Tamsin J., Fuller, Hannah V., Dasmahapatra, Kanchon K., & Coe, Rebecca L., 2004, Local traditions of bower decoration by spotted bowerbirds in a single population, *Animal Behaviour* 68 (4): 759–65.

Majerus, Mike, 2006, Book review: *The selfish meme, Cambridge: The Magazine of the Cambridge Society* 57: 29–30.

Majerus, Mike, 2007, *The peppered moth: The proof of Darwinian evolution*, talk given at the 55th Annual Meeting of the Entomological Society of Alberta in Uppsala, Sweden, August 23.

Majid, Asifa, Bowerman, Melissa, Kita, Sotaro, Haun, Daniel B. M., & Levinson, Stephen C., 2004, Can language restructure cognition? The case for space, *Trends in Cognitive Sciences* 8 (3): 108–14.

Makalowska, I., Lin, C. F., & Makalowski, W., 2005, Overlapping genes in vertebrate genomes, *Computational Biology and Chemistry* 29 (1): 1–12.

Mameli, Matteo, 2005, Book review: *The selfish meme, Notre Dame Philosophical Reviews*, September 16, http://ndpr.nd.edu/review.cfm?id=4001.

Marr, Andrew, 2004, *My trade: A short history of British journalism*, Macmillan.

Marsden, Paul, 1998, Memetics and social contagion: Two sides of the same coin? *Journal of Memetics – Evolutionary Models of Information Transmission* 2 (2), http://cfpm.org/jom-emit/1998/vol2/marsden_p.html.

Maynard Smith, J., & Szathmary, E., 1997, *The major transitions in evolution*, Oxford University Press.

Mayo, M. A., & Pringle, C. R., 1998, Virus taxonomy – 1997, *Journal of General Virology* 79: 649–57.

Mayr, Ernst, 1996, What is a species, and what is not? *Philosophy of Science* 63: 262–77.

McDonald, Kate, 2007, Epigenetic inheritance and colorectal cancer, *Life Scientist*, July 18, http://www.lifescientist.com.au/article/188882/epigenetic_inheritance_colorectal_cancer.

McDonald, M. A., Sigman, M., Espinosa, M. P., & Neumann, C. G., 1994, Impact of a temporary food shortage on children and their mothers, *Child Development* 65 (2): 404–15.

Meeker-O'Connell, Ann, 2001, *How photocopiers work*, HowStuffWorks.com, http://home.howstuffworks.com/photocopier.htm.

Meltzoff, Andrew N., 1988, Infant imitation after a 1-week delay: Long-term memory for novel acts and multiple stimuli, *Developmental Psychology* 24 (4): 470–6.

Meltzoff, Andrew N., 2005, Imitation and other minds: The "like me" hypothesis, in S. Hurley & N. Chater, eds., *Perspectives on imitation: From neuroscience to social science*, Massachusetts Institute of Technology Press, pp. 55–77.

Meltzoff, Andrew N., 2007, "Like me": A foundation for social cognition, *Developmental Science* 10 (1): 126–34.

Meltzoff, Andrew N., & Moore, M. Keith, 1994, Imitation, memory, and the representation of persons, *Infant Behaviour and Development* 17: 83–99.

Mesoudi, A., & Whiten, A., 2004, The hierarchical transformation of event knowledge in human cultural transmission, *Journal of Cognition and Culture* 4 (1): 1–24.

Mesoudi, Alex, 2007a, Biological and cultural evolution: Similar but different, *Biological Theory* 2 (2): 119–23.

Mesoudi, Alex, 2007b, The experimental study of cultural transmission and its potential for explaining archeological data, in M. J. O'Brien, ed., *Cultural transmission and archeology: Issues and case studies*, Society for American Archeology Press, pp. 91–101.

Mesoudi, Alex, 2007c, Extended evolutionary theory makes human culture more amenable to evolutionary analysis, *Behavioural and Brain Sciences* 30 (4): 374.

Mesoudi, Alex, 2008, Foresight in cultural evolution, *Biology and Philosophy* 23 (2): 243–55.

Mesoudi, Alex, Whiten, Andrew, & Dunbar, Robin, 2006, A bias for social information in human cultural transmission, *British Journal of Psychology* 97: 405–23.

Mesoudi, Alex, Whiten, Andrew, & Laland, Kevin N., 2004, Perspective: Is human cultural evolution Darwinian? Evidence reviewed from the perspective of *The Origin of Species*, *Evolution* 58 (1): 1–11.

Mesoudi, Alex, Whiten, Andrew, & Laland, Kevin N., 2006, Towards a unified science of cultural evolution, *Behavioral and Brain Sciences* 29: 329–83.

Miller, George A., 1956, The magical number seven, plus or minus two: Some limits on our capacity for processing information, *Psychological Review* 63: 81–97.

Millet, Kobe, & Dewitte, Siegfried, 2007a, Altruistic behavior as a costly signal of general intelligence, *Journal of Research in Personality* 41 (2): 316–26.

Millet, Kobe, & Dewitte, Siegfried, 2007b, Non-conformity: The hidden driver behind the positive relationship between IQ and vegetarianism? *BMJ Rapid Responses*, January 23. http://www.bmj.com/cgi/eletters/bmj.39030.675069.55v1#155201.

Mithen, Steven, 2000, Mind, brain and material culture: An archeological perspective in Peter Carruthers & Andrew Chamberlain, eds., *Evolution and the human mind: Modularity, language and meta-cognition*, Cambridge University Press, pp. 207–17.

Morgan, Drake, Grant, Kathleen A., Gage, H. Donald, Mach, Robert H., Kaplan, Jay R., Prioleau, Osric, Nader, Susan H., Buchheimer, Nancy, Ehrenkaufer, Richard L., & Nader, Michael A., 2002, Social dominance in monkeys: Dopamine D2 receptors and cocaine self-administration, *Nature Neuroscience* 5: 169–74.

Mulder, Moique Borgerhoff, Nunn, Charles L., & Towner, Mary C., 2006, Cultural macroevolution and the transmission of traits, *Evolutionary Anthropology* 15: 52–64.

Nakhleh, Luay, Warnow, Tandy, Ringe, Don, & Evans, Steven N., 2005, A comparison of phylogenetic reconstruction methods on an Indo-European dataset, *Transactions of the Philological Society* 103 (2): 171–92.

Nee, Sean, 2004, More than meets the eye, *Nature* 429: 804–5.

Needham, Rodney, 1975, Polythetic classification: Convergence and consequences, *Man* 10 (3): 349–69.

Nettle, David, & Dunbar, Robin I. M., 1997, Social markers and the evolution of reciprocal exchange, *Current Anthropology* 38: 93–9.

O'Brien, M. J., 2005, Evolutionism and North American's archaeological record, *World Archaeology* 37 (1): 26–45.

O'Brien, M. J., & Holland, T. D., 1995a, Behavioural archeology and the extended phenotype, in J. M. Skibo, W. H. Walker & A. E. Nielsen, eds., *Expanding archaeology*, University of Utah Press, pp. 143–61.

O'Brien, M. J., & Holland, T. D., 1995b, The nature and premise of a selection-based archaeology, in P. A. Teltser, ed., *Evolutionary archaeology: Methodological issues*, University of Arizona Press, pp. 175–200.

Odling-Smee F. J., Laland K. N., & Feldman, M. W., 2003, *Niche construction: The neglected process in evolution*, Monographs in Population Biology 37, Princeton University Press.

Offer, Avner, 1997, Between the gift and the market: The economy of regard, *Economic History Review* 3: 450–76.

Oyama, Susan, 2000, Causal democracy and causal contributions in developmental systems theory, *Philosophy of Science* 67: 332–47.

Paciotti, Brian, Richerson, Peter J., & Boyd, Robert, 2006, Cultural evolutionary theory: A synthetic theory for fragmented disciplines, in Paul Van Lange, ed., *Bridging social psychology: The benefits of transdisciplinary approaches*, Psychology Press, pp. 365–70.

Pagel, Mark, 2006, Darwinian cultural evolution rivals genetic evolution, *Behavioral and Brain Sciences* 29: 360.

Pagel, Mark, & Mace, Ruth, 2004, The cultural wealth of nations, *Nature* 248: 275–8.

Partridge, Linda, & Gems, David, 2002, The evolution of longevity, *Current Biology* 12: 544–6.

Pembrey, Marcus E., Bygren, Lars Olov, Kaati, Gunnar, Edvinsson, Sören, Northstone, Kate, Sjöström, Michael, Golding, Jean, & Whitelaw, Emma, 2006, Sex-specific, male-line transgenerational responses in humans, *European Journal of Human Genetics* 14 (2): 159–66.

Pepperberg, Irene Maxine, 1999, *The Alex studies: Cognitive and communicative abilities of grey parrots*, Harvard University Press.

Perry-Jenkins, Maureen, & Folk, Karen, 1994, Class, couples, and conflict: Effects of the division of labor on assessments of marriage in dual-earner families, *Journal of Marriage and the Family* 56 (1): 165–80.

Peterson, Ivars, 2006, From counting to writing, *Science News Online* 169 (10), http://www.sciencenews.org/view/generic/id/7121/title/From_Counting_to_Writing.

Petersson, Karl Magnus, Silva, Carla, Castro-Caldas, Alexandre, Ingvar, Martin, & Reis, Alexandra, 2007, Literacy: A cultural influence on functional left–right differences in the inferior parietal cortex, *European Journal of Neuroscience* 26: 791–9.

Petersson, Karl Magnus, Ingvar, Martin, & Reis, Alexandra, 2008, Language and literacy from a cognitive neuroscience perspective, in D. R. Olson & N. Torrence, eds., *Cambridge handbook of literacy*, Cambridge University Press, pp. 152–82.

Phillips, J. B., 1960, *God our contemporary*, Hodder and Stoughton.

Piattelli-Palmarini, Massimo, 1989, Evolution, selection and cognition: From "learning" to parameter setting in biology and the study of language, *Cognition* 31: 1–44.

Pinker, Steven, 1994, *The language instinct: The new science of language and mind*, Penguin Books.

Pinker, Steven, 2003, Language as an adaptation to the cognitive niche, in M. H. Christiansen & S. Kirby, eds., *Language evolution*, Oxford University Press, pp. 16–37.

Pinker, Steven, & Bloom, Paul, 1990, Natural language and natural selection, *Behavioral and Brain Sciences* 13 (4): 707–84.

Plotkin, H. C., 2002, *The imagined world made real*, Penguin.

Pratchett, Terry, Stewart, Ian, & Cohen, Jack, 2003, *The science of discworld II: The globe*, Ebury Press.

Price, If, 2004, Complexity, complicatedness and complexity: A new science behind organizational intervention? *Emergence: Complexity and Organization* 6 (1–2): 40–8.

Price, I., 2007, Lean assets: New language for new workplaces, *California Management Review* 49 (2): 102–18.

Price, I., 2008, Space to adapt: Workplaces, creative behaviour and organizational memetics, in Tudor Rickards, Susan Moger & Mark Runco, eds., *The Routledge companion to creativity*, Routledge, pp. 46–58.

Price, If, & Shaw, Ray, 1998, *Shifting the patterns: Breaching the memetic codes of corporate performance*, Management Books 2000.

Pyers, Jennie, 2006, Constructing the social mind: Language and false-belief understanding, in Nicholas J. Enfield & Stephen C. Levinson, eds., *Roots of human sociality: Culture, cognition and interaction*, Berg Publishers, pp. 207–28.

Pyper, Hugh S., 1997, *The selfish text: The Bible and memetics*, The Bible into Culture Colloquium, Sheffield, April 9–12.

Qiu, Jane, 2006, Epigenome: Unfinished symphony, *Nature* 441: 143–5.

Quiroga, R. Q., Reddy, L., Kreiman, G., Koch, C., & Fried, I., 2005, Invariant visual representation by single neurons in the human brain, *Nature* 435: 1102–7.

Railsback, L. Bruce, 2007, *A paean to prokaryotes, or "bully for bacillus,"* http://www.gly.uga.edu/railsback/11111misc/Prokaryotes.html.

Ramírez-Esparza, Nairán, Goslinga, Samuel D., Benet-Martínez, Verónica, Potter, Jeffrey P., & Pennebakera, James W., 2006, Do bilinguals have two personalities? A special case of cultural frame switching, *Journal of Research in Personality* 40: 99–120.

Range, F., Viranyi, Z., & Huber, L., 2007, Selective imitation in domestic dogs, *Current Biology* 17: 868–72.

Regier, Terry, & Kay, Paul, 2006, Language, thought and color: Recent developments, *Trends in Cognitive Sciences* 10 (2): 51–4.

Renzulli, Joseph S., 2005, The three-ring conception of giftedness: A developmental model for promoting creative productivity, in Robert J. Sternberg & Janet E. Davidson, eds., *Conceptions of giftedness*, 2nd ed.,Cambridge University Press, pp. 246–79.

Reynolds, Dorothy, 1998, *Giftedness in children between the ages of six and eleven years: A study of its relevance to the child, the family and the school*, unpublished master's thesis, Birkbeck College, London, and the Institute of Family Therapy.

Roach, Tom, 2006, Evolution in the head, *Campaign*, September 15, http://www.campaignlive.co.uk/news/rss/592967/IPA-Excellence–Diploma–Distinction–essayTom–Roach–AMV–BBDOEvolution–head/.

Robson, Julia, 2002, A burst of colour from Bollywood, *Daily Telegraph*, May 3.

Rodríguez Martínez, María del Carmen, Ortíz Ceballos, Ponciano, Coe, Michael D., Diehl, Richard A., Houston, Stephen D., Taube, Karl A., Calderó, & Alfredo Delgado, 2006, Oldest writing in the New World, *Science* 313: 1610.

Rogers, Alan, 1988, Does biology constrain culture? *American Anthropologist* 90 (4): 819–31.

Romney, A. Kimball, Moore, Carmella C., & Rusch, Craig D., 1997, Cultural universals: Measuring the semantic structure of emotion terms in English and Japanese, *Proceedings of the National Academy of Sciences of the United States of America* 94 (10): 5989–94.

Rose, Andrew K., 2000, Do currency unions increase trade? A "gravity" approach, *Federal Reserve Bank of San Francisco Economic Letter*, no. 3, http://www.frbsf.org/econrsrch/wklyltr/2000/el2000-03.html#subhead4.

Rose, Steven, 2005, What Darwin really thought, *Guardian*, July 23, http://www.guardian.co.uk/books/2005/jul/23/featuresreviews.guardianreview8.

Rose, Steven, 2007, Letter, *London Review of Books* 29 (22), http://www.lrb.co.uk/v29/n22/letters.html.

Rothschild, Michael, 1992, *Bionomics: The inevitability of capitalism*, Futura.

Rothwell, David, 2007, *The Wordsworth dictionary of homonyms*, Wordsworth Reference.

Rotman, Brian, 1987, *Signifying nothing: The semiotics of zero*, Palgrave Macmillan.

Saler, Benson, 1993, *Conceptualizing religion: Immanent anthropologists, transcendent natives and unbounded categories*, E. J. Brill.

Salomon, Gavriel, & Perkins, David N., 1998, Individual and social aspects of learning, *Review of Research in Education* 23: 1–24.

Schmandt-Besserat, Denise, 1996, *How writing came about*, University of Texas Press.

Schmandt-Besserat, Denise, 2002, Signs of life, *Odyssey* (January–February): 6.

Schmandt-Besserat, Denise, 2006, When writing met art: From symbols to narrative – summary document, https://webspace.utexas.edu/dsbay/Docs/WhenWritingMetArt.pdf.

Schofield, Jack, 2003, Why VHS was better than Betamax, *Guardian*, January 25, http://www.guardian.co.uk/technology/2003/jan/25/comment.comment.

Schumann, John, Favareau, Donald, Goodwin, Charles, Lee, Namhee, Mikesell, Lisa, Tao, Hongyin, Véronique, Daniel, & Wray, Alison, 2006, Language evolution: What evolved? *Saint-Chamas: Marges Linguistiques* 11: 167–99.

Schwartz, Michael Alan, & Wiggins, Osborne P., 1987, Diagnosis and ideal types: A contribution to psychiatric classification, *Comprehensive Psychiatry* 28 (4): 277–91.

Seeger, Charles, 1958, Prescriptive and descriptive music-writing, *Musical Quarterly* 44 (2): 184–95.

Shu, Hua, & Anderson, Richard C., 1999, Learning to read Chinese: The development of metalinguistic awareness, in Jian Wang, Albrecht W. Inhoff & Hsuan-Chih Chen, eds., *Reading Chinese script: A cognitive analysis*, Lawrence Erlbaum Associates, pp. 1–18.

Simmel, Georg, 1978, *The Philosophy of Money*, Routledge.

Simmons, David, & Williams, Rhys, 1997, Dietary practices among Europeans and different South Asian groups in Coventry, *British Journal of Nutrition* 78: 5–14.

Singh, Supriya, 2000, Electronic commerce and the sociology of money, *Sociological Research Online* 4 (4), http://www.socresonline.org.uk/4/4/singh.html.

Singh, S., & Slegers, C., 1997, *Trust and electronic money*, Centre for International Research on Communication and Information Technologies Policy Research Paper 42.

Smith, Adam, 1776, *An inquiry into the nature and causes of the wealth of nations*, Oxford University Press.

Smith, Kenny, Brighton, Henry, & Kirby, Simon, 2003, Complex systems in language evolution: The cultural emergence of compositional structure, *Advances in Complex Systems* 6 (4): 547–58.

Snow, C. P., 1969, *The two cultures and a second look: An expanded version of the two cultures and the scientific revolution*, Cambridge University Press.

Sober, Elliott, 1993, *Philosophy of biology*, Oxford University Press.

Sperber, Dan, 1996, *Explaining culture: A naturalistic approach*, Blackwell.

Sperber, Dan, 2007, Seedless grapes: Nature and culture, in Stephen Laurence & Eric Margolis, eds., *Creations of the mind: Theories of artifacts and their representation*, Oxford University Press, pp. 124–37.

Sperber, Dan, 2000a, Metarepresentations in an evolutionary perspective, in *Metarepresentation: A multidisciplinary perspective*, Oxford University Press, pp. 117–37.

Sperber, Dan, 2000b, An objection to the memetic approach to culture, in Robert Aunger, ed., *Darwinizing Culture*, Oxford University Press, pp. 163–74.

Spierdijk, Laura, & Vellekoop, Michel, 2009, The structure of bias in peer voting systems: Lessons from the Eurovision Song Contest, *Empirical Economics* 36: 403–25, http://www.springerlink.com/content/r52473276n234705/fulltext.pdf.

Stanovich, K. E., & West, R. F., 2003, Evolutionary versus instrumental goals: How evolutionary psychology misconceives human rationality, in D. E. Over, ed., *Evolution and the psychology of thinking: The debate*, Psychological Press, pp. 171–230.

Stanovich, Keith E., 2005, *The robot's rebellion: Finding meaning in the age of Darwin*, University of Chicago Press.

Stanovich, Keith E., 2006, Memetics and money, *Behavioural and Brain Sciences* 29: 194–5.

Stanovich, Keith E., & West, Richard F., 2007, Natural myside bias is independent of cognitive ability, *Thinking and Reasoning* 13 (3): 225–47.

Steels, L., 1999, The puzzle of language evolution, *Kognitionswissenschaft* 8 (4): 143–50.

Sterelny, Kim, 2001, *The evolution of agency and other essays*, Cambridge University Press.

Sterelny, Kim, 2004a, Externalism, epistemic artefacts and the extended mind, in Richard Schant, ed., *The externalist challenge*, de Gruyter, pp. 239–54.

Sterelny, Kim, 2004b, Review: Genes, memes and human history by Stephen Shennan, *Mind and Language* 19 (2): 249–57.

Sterelny, Kim, 2006, Memes revisited, *British Journal for the Philosophy of Science* 57: 145–65.

Sterelny, Kim, Smith, Kelly, & Dickison, Mike, 1996, The extended replicator, *Biology and Philosophy* 11 (3): 377–403.

Szamado, Szabolcs, & Szathmary, Eors, 2006, Selective scenarios for the emergence of natural language, *Trends in Ecology and Evolution* 21 (10): 555–61.

Szathmáry, Eors, 2000, The evolution of replicators, *Philosophical Transactions of the Royal Society* B 355 (1403): 1669–76.

Tehrani, Jamshid, & Collard, Mark, 2002, Investigating cultural evolution through biological phylogenetic analyses of Turkmen textiles, *Journal of Anthropological Archaeology* 21 (4): 443–63.

Tëmkin, Ilya, & Eldredge, Niles, 2007, Phylogenetics and material cultural evolution, *Current Anthropology* 48: 146–53.

Templeton, A. R., 1989, The meaning of species and speciation: a genetic perspective, in D. Otte & J. Endler, eds., *Speciation and its consequences*, Sinauer, pp. 3–27.

Templeton, A. R., Sing, C. F., & Brokaw, B., 1976, The unit of selection in *Drosophila mercatorium* 1. The interaction of selection and meiosis in parthenogenetic strains, *Genetics* 82: 249–376.

Tenreyro, Silvana, & Barro, Robert J., 2003, *Economic effects of currency unions*, National Bureau of Economic Research Working Paper No. 9435, January.

Tomasello, Michael, 2006, Why don't apes point? in Nicholas J. Enfield & Stephen C. Levinson, eds., *Roots of human sociality: Culture, cognition and interaction*, Berg Publishers.

Tomasello, Michael, Carpenter, Malinda, Call, Josep, Behne, Tanya, & Moll, Henrike, 2005, Understanding and sharing intentions: The origins of cultural cognition, *Behavioral and Brain Sciences* 28: 675–735.

Tomasello, Michael, Carpenter, Malinda, & Liszkowski, Ulf, 2007, A new look at infant pointing, *Child Development* 78 (3): 705–22.

Tooby, J., & Cosmides, L., 1992, The psychological foundations of culture, in J. Barkow, L. Cosmides & J. Tooby, eds., *The adapted mind*, Oxford University Press, pp. 19–136.

Toplak, Maggie E., & Stanovich, Keith E., 2003, Associations between myside bias on an informal reasoning task and amount of post–secondary education, *Applied Cognitive Psychology* 17: 851–60.

Torres, Susan J., 2007, Relationship between stress, eating behavior, and obesity, *Nutrition* 23 (11–12): 887–94.

Truss, Lynn, 2003, *Eats shoots and leaves: The zero tolerance approach to punctuation*, Profile Books.

Turgenev, Ivan, 1861, *Fathers and sons*, Penguin Books, 1975.

Tversky, A., & Kahneman, D., 1983, Extensional versus intuitive reasoning: The conjunction fallacy in probability judgment, *Psychological Review* 90: 293–315.

Tyler, Michael Douglas, 2003, Orthography, phonology, and the measurement of vocal response times, unpublished Ph.D. dissertation, University of West Sydney, Sydney, Australia.

Van Coetsem, Frans, 1988, *Loan phonology and the two transfer types in language contact*, Foris.

Van Coetsem, Frans, 2000, *A general and unified theory of the transmission process in language contact*, Carl Winter.

Van Driem, George, 2007, *Symbiosism, symbiomism and the Leiden definition of the meme*, http://www.semioticon.com/virtuals/imitation/van_driem_paper.pdf.

Van Regenmortel, M. H., 1990, Virus species, a much overlooked but essential concept in virus classification, *Intervirology* 31: 241–54.

Van Regenmortel, Marc H. V., 2003, Viruses are real, virus species are man-made, taxonomic constructions, *Archives of Virology* 148: 2481–8.

Van Regenmortel, Marc H. V., & Mahy, Brian W. J., 2004, Emerging issues in virus taxonomy, *Emerging Infectious Diseases*, January, http://www.cdc.gov/ncidod/EID/vol10no1/03-0279.htm.

Van Regenmortel, M. H. V., Bishop, D. H. L., Fauquet, C. M., Mayo, M. A., Maniloff, J., & Calisher, C. H., 1997, Guidelines to the demarcation of virus species, *Archives of Virology* 142: 1505–18.

Vaneechoutte, Mario, & Skoyles, John R., 1998, The memetic origin of language: Modern humans as musical primates, *Journal of Memetics – Evolutionary Models of Information Transmission* 2 (2), http://cfpm.org/jom-emit/1998/vol2/vaneechoutte_m&skoyles_jr.html.

Verhaegen, Marc, 1988, Aquatic ape theory, speech origins: A hypothesis, *Speculations in Science and Technology* 11: 165–71.

Vohs, Kathleen D., Mead, Nicole L., & Goode, Miranda R., 2006, The psychological consequences of money, *Science* 314: 1154–6.

Vohs, Kathleen D., Mead, Nicole L., & Goode, Miranda R., 2008, Merely activating the concept of money changes personal and interpersonal behavior, *Current Directions in Psychological Science* 17 (3): 208–12.

Vygotsky, L. S., 1986, *Thought and language*, Massachusetts Institute of Technology Press.

Warnow, Tandy, Evans, Steven N., Ringe, Don, & Nakhleh, Luay, 2004, A stochastic model of language evolution that incorporates homoplasy and borrowing, in Colin Renfrew, Peter Forster & James Clackson, eds., *Phylogenetic methods and the prehistory of languages*, McDonald Institute for Archaeological Research, pp. 75–87.

Wason, P. C., 1966, Reasoning, in B. Foss, ed., *New horizons in psychology*, Penguin, pp. 135–51.

Weaver, Warren, & Shannon, Calude, 1949, eds., *The mathematical theory of communication*, University of Illinois Press.

Weinzirl, John, & Gerdeman, Alice E., 1929, The bacterial count of ice cream held at freezing temperatures, *Journal of Dairy Science* 12 (2): 182–9.

Wendt, Dirk, 2008, Book review: The selfish meme, *Journal of Biosocial Science* 40: 157–60.

Werker, J. F., & Lalonde, C. E., 1988, Cross-language speech perception: Initial capabilities and developmental change, *Developmental Psychology* 24: 672–83.

White, Tim, 2003, *Interview by Harry Kreisler*, Conversations with History: Institute of International Studies, University of California, Berkeley, http://globetrotter. berkeley.edu/people3/White/white-cono.html.

Whiten, A., & Byrne, R., 1988, Tactical deception in primates, *Behavioural and Brain Sciences* 12: 233–73.

Whiten, A., Horner, V., & deWaal, F. B., 2005, Conformity to cultural norms of tool use in chimpanzees, *Nature* 437: 737–40.

Whiten, Andrew, 2001, The evolutionary roots of culture, *British Academy Review* 6: 57–60.

Whiten, Andrew, 2005, The second inheritance system of chimpanzees and humans, *Nature* 437: 52–5.

Whitman, William B., Coleman, David C., & Wiebe, William J., 1998, Prokaryotes: The unseen majority, *Proceedings of the National Academy of Sciences* 95: 6578–83.

Wilkie, J. R., Ferree, M. M., & Ratcliff, K. S., 1998, Gender and fairness: Marital satisfaction in two-earner couples, *Journal of Marriage and the Family* 60: 577–94.

Wilkins, John S., 2006a, The concept and causes of microbial species, *Studies in History and Philosophy of the Life Sciences* 28: 389–408.

Wilkins, John S., 2006b, Species, kinds, and evolution, *Reports of the National Center for Science Education* 26 (4): 36–45.

Wilkins, John S., 2007, The dimensions, modes and definitions of species and speciation, *Biology and Philosophy* 22: 247–66.

Wilkins, John S., 2008, The adaptive landscape of science, *Biology and Philosophy* 23 (5): 659–71.

Wilkins, John S., 2009, *Myth 3: Darwin was a Lamarckian*, http://scienceblogs. com/evolvingthoughts/2009/02/myth_3_darwin_was_a_lamarckian.php.

Wilson, Mark E., Fisher, Jeff, Fischer, Andrew, Lee, Vannessa, Harris, Ruth B., & Bartness, Timothy J., 2008, Quantifying food intake in socially housed monkeys: Social status effects on caloric consumption, *Physiology and Behavior* 94 (4): 586–94.

Winford, Donald, 2003, *An introduction to contact linguistics*, Blackwell.

Winford, Donald, 2005, Contact-induced changes: Classification and processes, *Diachronica* 22 (2): 373–427.

Wittgenstein, Ludwig, 1953, *Philosophical investigations*, Wiley Blackwell.

Wood, J., Hennell, T., Jones, A., Hooper, J., Tocque, K., & Bellis, M. A., 2006, *Where wealth means health: Illustrating inequality in the North West*, North West Public Health Observatory.

Xing, Y., Shi, S., Le, L., Lee, C. A., Silver-Morse, L., & Li, Willis X., 2007, Evidence for transgenerational transmission of epigenetic tumor susceptibility in Drosophila, *Public Library of Science Genetics* 3 (9): e151.

Yair, Gad, 1995, "Unite Unite Europe": The political and cultural structures of Europe as reflected in the Eurovision Song Contest, *Social Networks* 17 (2): 147–61.

Yair, Gad, & Maman, Daniel, 1996, The persistent structure of hegemony in the Eurovision Song Contest, *Acta Sociologica* 39: 309–25.

Yehuda, Rachel, Bell, Amanda, Bierer, Linda M., & Schmeidler, James, 2008, Maternal, not paternal, PTSD is related to increased risk for PTSD in offspring of Holocaust survivors, *Journal of Psychiatric Research* 42 (13): 1104–11.

Yehuda, Rachel, Bierer, Linda M., Schmeidler, James, Aferiat, Daniel H., Breslau, Ilana, & Dolan, Susan, 2000, Low cortisol and risk for PTSD in adult offspring of holocaust survivors, *American Journal of Psychiatry* 157: 1252–9.

Yehuda, Rachel, Halligan, Sarah L., & Bierer, Linda M., 2002, Cortisol levels in adult offspring of Holocaust survivors: Relation to PTSD symptom severity in the parent and child, *Psychoneuroendocrinology* 27 (1–2): 171–80.

Yehuda, Rachel, Mulherin Engel, Stephanie, Brand, Sarah R., Seckl, Jonathan, Marcus, Sue M., & Berkowitz, Gertrud S., 2005, Transgenerational effects of posttraumatic stress disorder in babies of mothers exposed to the World Trade Center attacks during pregnancy, *Journal of Clinical Endocrinology and Metabolism* 90 (7): 4115–8.

Young, H. P., & Burke, M. A., 1998, Competition and custom in economic contracts: A case study of Illinois agriculture, *American Economic Review* 93 (3): 559–73.

Zelizer, Viviana, 2005, Missing monies: Comment on Nigel Dodd, "Reinventing monies in Europe," *Economy and Society* 43 (4): 584–8.

Zellner, Debra A., Loaiza, Susan, Gonzalez, Zuleyma, Pita, Jaclyn, Morales, Janira, Pecora, Deanna, & Wolf, Amanda, 2006, Food selection changes under stress, *Physiology and Behavior* 87 (4): 789–93.

Zheng, Yujian, 2008, *Memes, mind and normativity*, Proceedings of the International Conference on the Theme of Cognition and the Study of Culture, May 29–31, Center for the Study of Language and Cognition, Zhejiang University, Hangzhou, China.

Zhou, Xinyue, Vohs, Kathleen D., & Baumeister, Roy F., 2009, The symbolic power of money: Reminders of money alter social distress and physical pain, *Psychological Science* 20 (6): 700–6.

Zuberbühler, K., Cheney, D. L., & Seyfarth, R. M., 1999, Conceptual semantics in a non-human primate, *Journal of Comparative Psychology* 113 (1): 33–42.

Index

abstract numbers, 100
Adler, Nancy, 160
affect labelling, 79
alarm calls, 12, 14–15, 20, 50
Alexander, Ruth, 157
Altbäcker, V., 33, 34
American English, 115, 117, 118
Andreosso-O'Callaghan, Bernadette, 122, 123
apes. *See* nonhuman primates
Aquatic Ape Hypothesis, 56
Aronoff, Stan, 195
artefactual language. *See also* computer programming languages; currency; mathematical notation; money; musical notation; orthographic projection; writing
   accuracy, 113–114, 119–120
   advantages, 6
   badge of representational identity, 118
   biological fitness, 125
   capacity (scaffolding), 111–113
   capacity (swap space), 110–111
   coevolves with content and medium, 101
   communicative role, 102
   compositionality, 99
   cooperative game, 118, 124
   cultural artefact, 90–92
   definition, 6, 49
   detachment, 120–122
   dialects, 117–120
   emergence, 5–6
   evolved for representation, 89, 92, 95, 100, 116–117, 120, 225

   facilitates functional cooperation, 121, 122, 123–125, 225
   homonymy, 116
   longevity and fecundity, 107–110
   poem example, 94
   significance for culture, 223
   social status, 124
   synonymy, 113–114
associative learning, 13–14
Aunger, Robert, 94
Aureli, F., 164
autism, 61, 226

Ball, Andrew, 204
Bapteste, Eric, 201, 202, 203
Barclay, Robert, 163–164
Baron-Cohen, Simon, 181, 226
barter
   based on trust, 127, 132
   bilateral, 128, 131, 132
   definition, 127
   limitations, 127–128, 130
   noncompositional, 145
Bateson, Gregory, 25
bats, 163
beavers, 30–32
Bekkering, Harold, 52–53
Benet-Martínez, Verónica, 82–84
Berk, L., 112
Betamax. *See* technology
Bilkó, A., 33, 34
biological taxonomy
   evaluation, 215
   organisms versus genes, 201–202
   prokaryotes, 202–203
   sexual organisms, 200–201

biological taxonomy (*cont.*)
  species versus ecological clusters,
    204–206
  summary, 208
  viruses, 203–204
Bloom, Paul, 70–71
Bock, Walter, 200, 204–205, 216
Bogoshi, Jonas, 95
Booth, Alan, 153
borderline personality disorder, 226
bottlenecks in language learning, 74, 76,
  95, 99, 126
Boyd, Robert
  cognitive attractors, 43–44
  conformity, 172
  discrete replicators not necessary for
    evolution, 40, 45
  Illinois sharecroppers, 41–43, 44
brain (human)
  effects of literacy, 103–104
  postnatal development, 65
  size and proportions, 56, 67
  stores information hierarchically, 42
Brighton, Henry, 74
Brockwood, Krista, 153
Bruner, J., 149
Burke, M., 44
Byrne, Richard, 52, 57

Call, Josep and Carpenter, Malinda, 53
Cascajal block, 109
causal complexity, 23–24
cellular mechanisms, 26
Chalmers, David, 112
Chandler, Daniel, 13
Chaney, Richard, 198, 199
Cheney, D., 14
chimps. *See* nonhuman primates
Chiu, Chi-yue, 82–84
Chomsky, Noam, 70, 72, 79
Clark, Andy, 23–24, 112–113
clay tokens
  alternatives, 100
  coin evolution analogy, 138–139
  earliest forms, 96–97
  marks on envelopes, 98
  marks on tablets, 98–99
  replaced by stylus marks, 99
  representational advantages, 97–98
  representational disadvantages, 97, 99
  sealed in clay envelopes, 97

Clay, Rebecca, 160
Clerides, Sofronis, 156
Cloak, Ted, 187, 195
Coetsem, Frans Van, 84
cognitive attractors, 44
Cohen, Jack, 20
Cohen, Sheldon, 160
coins, 129, 134, 138–139
Coleman, David, 202
Collard, Mark, 218–219, 227
colour perception, 79–80
communication
  and representation in speech, 104–106
  dependent on cooperation, 58, 76
  dependent on shared language, 67
  fitness effects, 62, 63
  Gricean, 76, 174–175, 223
  hierarchical in humans, 105–106
  instinctive in humans, 63–64
  interpretation (distinct from), 66–67,
    76, 119
  representation (distinct from), 67–69
  writing, 101–104
computer programming languages, 6, 113
concrete numbers, 100
Connor, R., 164
contact linguistics, 84–85, 104
cooperation
  dependent on relatedness, 64
  functional versus social, 121–122,
    123–125, 224–226
  human characteristic, 58
  pointing, 59
  preadaptation for natural language,
    57–60
critical period, 64–66, 69, 74, 76, 86
Crosson, B., 42
Csibra, Gergely, 63–64
cui bono?, 3, 148
cultural agents, 169, 180–182, 188–191,
  192, 199, 226, 229, 232
cultural ecology. *See also* cultural taxonomy
  environmental factors, 191–192
  metarepresentation spectrum, 206–208
  skills, 124, 180–181, 188–191
cultural evolution
  analogy with biology, 4–5, 109–110
  explanatory value, 227–229
  kick-started by natural language, 69
  metarepresentation, 95
  summary, 4, 126–127, 185, 229

cultural information
  distinct from cultural languages, 39
  effects, 187
  expression, 186–191
  hierarchically stored, 42
  hierarchically transmitted, 105–106
  money, 138
  printing technology, 108–109
  role in culture, 49
  transmission mechanisms, 206–207
cultural priming, 82–84, 180
cultural taxonomy. *See also* cultural ecology
  anecdotal, 196–197
  classification decisions, 216–218
  evaluation, 215–216
  isolation mechanisms, 212–214
  phylogenies, 208–209
  polythetic classes, 197–199
  reticulate, 203
  species barriers, 207–208
  summary, 7
  varying patterns, 218–219, 226–227
cultural units, 186–188
culture
  shaped by natural language, 82–84
  uniquely human, 5
currency. *See also* money
  artefactual language, 136–137, 144–145
  cultural evolution, 142–144
  debt measurement, 131
  different forms, 140–144
  euro, 143–144
  trade, 146–148, 156
Curtis, Tom, 202
cytoplasm, 5, 26

Damerow, Peter, 96, 100, 228
Darwin, Charles, 4, 112
Davies, Glyn, 128
Davies, Ron, 135
Dawkins, Richard
  "copy me," 233
  extended phenotype, 31, 201
  memes, 231
  taxonomy, 217
Deacon, Terrence
  human brain, 56
  replicas versus replicators, 26, 233
  representational systems, 20
  representations, 50
Dennett, Daniel, 3

developmental inheritance, 33–34
Developmental Systems Theory, 23, 221
Devlin, Keith, 20
deWaal, F. B., 51
Dewitte, Siegfried, 172, 177
Dickins, Benjamin, 36
Dickins, Thomas, 32, 36
Dietrich, Eric and Markman, Arthur,
    18–19, 23, 38, 41, 42, 44
discrete representations, 18–19, 40–45.
  See also receivers
  Illinois sharecroppers example, 41–42,
    44
  support from cognitive
    neuropsychology, 42
DNA
  analogous to natural language, 77
  epigenetics, 28, 29–30
  information, 12, 13, 21
  inheritance mechanism, 30
  inherited resource, 23, 221
  persistent heredity, 222
  replication and expression, 26–28
  stable structure, 107
  unambiguous, 113
  viral, 5
Dodd, Nigel, 130, 131, 143–144
dogs, 54–55
dolphins, 164
Doolittle, W. Ford, 201, 202, 203
Dunbar, Robin
  critical period, 65
  human bias for social information, 104
  language as a social badge, 64–65
  language as grooming, 62
  language evolution, 57
  social brain hypothesis, 57, 58
Dupoux, Emmanuel, 64
Durkheim, Émile, 122

Easterlin, Richard, 122
ecological inheritance, 30–32, 36–37
education
  cultural priming, 180
  economics, 122–123
  metarepresentation, 179–180, 207
  myside bias, 179
Eerkens, Jelmer, 113
email, 102, 141
emotions, 78–79
Enfield, Nicholas, 53, 58, 60

Epel, Elissa, 161
epigenetics, 28–30, 36, 222
euro, 143–144
Eurovision, 155–159
extended phenotype, 31, 201
eye colour, 23

Feldman, M., 30–31
Ferree, M., 153
fission-fusion societies, 58, 163–164
Folk, Karen, 153
Frankel, Jeffrey, 147

Garvin, R., 112
Gatherer, Derek, 157, 158–159
genes
    beavers and dams, 31
    causal role, 21, 22, 23–24, 221–222
    expression, 186
    fixed, 1, 2
    human intelligence, 173
    human rationality, 173–175
    ignored by humans, 3
    information, 2, 11
    inheritance mechanism, 26–28
    replicas not replicators, 26, 233
    role in culture, 3, 79, 221
    species barriers, 191, 204–206
    taxonomy, 201–203
    viral taxonomy, 203–204
Gerdeman, Alice, 202
Gergely, György, 52–53, 63–64
giftedness. *See also* Renzulli, Joseph
    adolescence, 163
    artefactual language use, 175–176
    fitness, 173–177
    metarepresentation, 7, 170–171
    nonconformism, 172–173, 177
gifts, 153–155
Giles, H., 65
Ginsburgh, Victor, 156
Glick, Reuven, 147–148, 156
Godfrey-Smith, Peter, 21
Goodman, C., 149
Goodwin, Charles, 59, 67
Gould, Stephen Jay, 70
Gray, Russell, 203
Greenhill, Simon, 203
Gricean communication. *See*
    communication
Griffiths, Paul, 162

Hall, William, 108–109
Henrich, Joseph
    cognitive attractors, 43–44
    discrete replicators not necessary for
        evolution, 40, 45
    Illinois sharecroppers, 41–43, 44
heredity. *See* inheritance
hierarchies
    in biological taxonomy, 200, 201, 202,
        203
    in cultural taxonomy, 218
    in human communication, 105–106
    in imitation, 52, 75
    in natural language classification, 186,
        199
    in nonhuman primates, 160
    in protocuneiform texts, 96, 228
    in representational systems, 77
    of information in human brains, 42
homonyms, 66
Hong, Ying-Yi, 82–84, 180
Horner, V., 51
Howe, Carl, 194
Huber, Ludwig, 54
Hudson, R., 33, 34
Hull, David, 162–163
humans. *See also* brain (human); critical
        period; cultural agents;
        metarepresentation
    able to ignore genes, 3
    Aquatic Ape Hypothesis, 56
    conformity bias, 65, 75, 76, 172
    cooperative, 58–60
    discretizing cognition, 42
    diversity, 180–181
    fossil record, 56
    hierarchical communication, 105–106
    hierarchical imitation, 52, 75
    instinctively communicative, 63–64
    language acquisition, 38
    pointing, 59
    selective imitation, 52, 53–54, 75
    social brain hypothesis, 58
    theory of mind, 60
    unique level of culture, 2
Hurford, James
    communication versus interpretation,
        66
    communication versus representation,
        67, 68–69
    critical period, 65

cultural evolution of language, 70,
72–73
evolution of grammar, 63
language and autism, 61
language as cooperation, 55, 58, 174
nonhuman representations, 50, 51
origins of language, 50
primate sociality, 58, 59
ritualization in animal behaviour, 14, 15,
50

icon, 50
Illinois sharecroppers. *See* discrete
representations
imitation
cognitive skills, 51
different levels, 52
dogs experiment, 54–55
lightbox experiments, 52–53
preadaptation for natural language,
51–55, 75
selective, 52–55
index, 50
information. *See also* cultural information;
receivers
associative learning, 17
birdsong example, 14, 19
clock example, 13, 17, 22
definition, 5, 25
depends on receiver, 16–18, 20, 24, 25
discretized by human cognition, 42–43
discretized by language, 43
DNA, 21
flowering plants example, 15–16, 19
heritability, 5, 12
Jablonka and Lamb, 12
monkeys example, 14–15, 19, 20
postal metaphor, 13
prescriptive versus descriptive, 21–23
recipe example, 12, 17, 21, 22
restaurant example, 13, 19, 22
rotary switch example, 16, 19
Shannon-Weaver model, 12
train timetable example, 21, 22
weather example, 13–14, 19
information theory, 12–13
Ingham, Geoffrey
barter, 132
issuing authorities, 131, 132, 145
money as a measure of value, 128, 132,
133, 145

money is not a commodity, 130–131
Ingvar, Martin, 91
inheritance
behavioural, 34–37
biological, 221–222
cultural, 39–40, 206–207, 222–223
definition, 12
developmental, 33–34, 36–37
ecological, 30–32, 36–37
epigenetic, 28–30, 36
genetic, 26–28, 36
genetically mediated, 36–37
linguistic, 37–39
of representational knowledge, 25
intelligence. *See* giftedness
Irwin, Rebecca, 110
Izak, G., 150–151

Jablonka, Eva and Lamb, Marion
body-to-body inheritance, 33
cultural inheritance, 37
four-dimensional model of evolution,
12, 25
information, 13
receivers, 12–13, 17, 18
socially mediated learning, 35
James, P., 161
Jessner, Ulrike, 86
Johnson, Boris, 162
Jordan, Peter, 218, 227
Jorion, Paul, 148, 161

Kaplan, Audrey, 195
Kasser, T., 152
Kay, Paul, 79–80
Khosla, Vipul, 228
Kinzler, Katherine, 64
Király, Ildikó, 52–53
Kirby, Simon, 49, 70, 74, 76–77
Kuhl, Patricia, 71–72, 81
Kurta, Alan, 163–164

Laland, K., 30–31, 208
Lalonde, C., 72
Lamb, Marion. *See* Jablonka, Eva and
Lamb, Marion
language. *See also* artefactual language;
language and cognition; natural
language
coevolves with content and medium,
101, 117

language (*cont.*)
  compositionality, 73
  cooperative game, 107
  dependent on cooperation, 58
  discretizes its content, 43, 80
  inherited resource, 37
  levels of precision, 116–117
  prepares receivers for information, 40
  synonymy and homonymy, 66, 113
language acquisition device, 70
language and cognition
  artefactual language, 93
  colour, 79–80
  culture, 82–84
  emotions, 78–79
  language learning, 80–81
  spatial cognition, 81–82
  speech perception, 71, 81
Lapavitsas, Costas
  monetary media, 129–130, 131
  money as a social nexus, 128, 131, 133, 145, 152
  money as universal commodity, 128–129, 131, 132, 140
Lea, Stephen and Webley, Paul, 148, 149–151, 153
Lebombo bone, 95
Lee, Ji-Hyun, 229
Leiser, D., 150–151
Levinson, Stephen
  cooperation, 58
  imitation, 53
  language, 71, 80
  theory of mind, 60
Lewis-Williams, J. D., 95
Li, Norman, 177
Lieberman, Matthew, 79
lightbox experiments, 52–53, 63–64
Linda problem, 174–175
Lipo, Carl, 113
Liszkowski, Ulf, 60
literacy
  brain structures, 91
  economics, 122–123
  IQ, 179
  metarepresentation, 103–104
  printing technology, 109
  self-perception, 162
Livesley, W., 198

Mace, Ruth, 147, 191
Mahy, Brian, 197, 203–204

Majid, Asifa, 81–82
mandolin, 190–191
maps, 112
Markman, Arthur. *See* Dietrich, Eric and Markman, Arthur
mathematical notation
  abstract versus concrete numbers, 100
  dialects, 117
  divergence from writing, 99–100
  Lebombo bone, 95
  representational efficiency, 93
  zero, 94
Mayo, M., 204
McDonald, Kate, 29
medium. *See* representational medium
Meltzoff, Andrew
  infant imitation, 55
  infant imitation and theory of mind, 60
  language learning, 71, 81
  lightbox experiments, 52
Mesoudi, Alex
  cultural taxonomy, 208
  flexible learning, 34
  human bias for hierarchical communication, 105–106
  human bias for social information, 104
  human reliance on social learning, 172
metarepresentation
  artefactual language, 175–176
  cultural transmission, 171–173
  definition, 6, 86
  education, 179–180
  frees information from particular languages, 7, 93–95, 224
  giftedness, 7, 170–171
  protowriting, 98
  significance for cultural evolution, 95
  spectrum in humans, 170–171
Miller, George, 43
Millet, Kobe, 172, 177
Mithen, Steven, 57, 91
mobile phones. *See* technology
money. *See also* barter; coins; currency
  barter (qualitatively different), 127–128
  biological basis, 148–149
  cattle, 129, 139, 140, 143
  Chinese, 134
  commodity, 128–133, 136, 141, 145
  credit, 133–135, 137, 140, 141, 143

cultural evolution, 137–140
 derivation of value, 131, 132–133,
  136–137
 different forms, 140–144
 Egyptian, 134
 electronic, 141–142
 Eurovision, 155–159
 fiat, 135–137, 140, 141, 143
 functional relationship, 136–137
 gift, 153–155
 grain, 129, 130, 134, 138, 140, 143
 imaginary, 135–137
 measure of value, 130–131, 133, 137,
  138, 139–140, 141, 144, 148
 media, 129, 133–135, 138, 139–144
 relationships, 151–153
 Roman, 134
 size illusion, 149–151
 social nexus, 128
 status, 159–165
 three-fold analysis, 128, 137, 144–145
 writing (evolutionary analogy), 138–139
monkeys. *See* nonhuman primates
Moore, C. C., 79
Moore, M. Keith, 55
Morgan, Drake, 160
Morris, Michael, 82–84, 180
musical notation. *See also* Seeger, Charles
 ambiguity, 116
 coevolved with content, 116–117, 213
 cooperative game, 115
 descriptive, 115
 facilitates functional cooperation, 124
 information, 190
 precision, 114, 121
 prescriptive, 114–117
 staff notation, 119
myside bias, 179–180

Naido, Kevin, 95
narrative, 104–106
natural language. *See also* communication;
  contact linguistics; critical period;
  language and cognition
 analogous to DNA, 77
 badge of group identity, 64–65, 75, 119,
  225
 biologically or culturally evolved?,
  69–73
 communication (evolutionary
  imperative), 62–63, 64, 69, 72–73,
  75–76, 78, 89, 90, 224

compositionality, 73–74
cooperative game, 59, 76
cultural artefact, 73, 76, 77
cultural prime, 82–84
facilitates social cooperation, 121
fitness, 63
grooming, 62
homonymy, 69
imitation as a preadaptation, 51–55
inherited resource, 38
listemes, 73
physiology as a preadaptation, 55–57
recodes information (Miller), 43
representation as a preadaptation,
  50–51
scaffolding, 112
second and subsequent languages, 65,
  85
significance for culture, 5, 69, 76–77,
  89, 223
sociality as a preadaptation, 57–60
sound-based medium, 68
taxonomy, 147
theory of mind as a preadaptation,
  60–61
universal grammar, 70, 71, 72
nature versus nurture, 3–4, 177–179
Nee, Sean, 202
Needham, Rodney, 197
Nettle, David, 64–65
Nicaraguan Sign Language, 60
niche construction, 30
noncompositional representations, 73, 95,
  97, 145
nonhuman culture, 2–3, 224
nonhuman primates. *See also* fission-fusion
  societies
 alarm calls, 50
 conformity, 51
 hand control, 56
 imitation, 51, 52–53
 pointing, 59
 representations, 50
 social hierarchy, 160
 social status and health, 160–161
 sociality, 57–58
Noury, Abdul, 156

Odling-Smee, F., 30–31
Offer, Avner, 152, 153, 155
orthographic projection, 117–118,
  120

Pagel, Mark, 147, 191
Pembrey, Marcus, 28
Penfield, Wilder, 56
Perkins, David, 189
Perry-Jenkins, Maureen, 153
Peterson, Ivars, 99
Petersson, Karl Magnus, 91, 92, 103–104
physiology (language preadaptation), 55–57
Pinker, Steven, 43, 70–71, 73
polythetic classes, 197–199, 204
Pratchett, Terry, 20
Price, If
  cultural evolution, 229
  culture and workspace, 191, 228
  economy of artefactual languages, 111
  language as communication, 63
  language shapes thought, 79, 94
  manufacturing plant design, 195
  nature and nurture, 178
  pattern shifting, 172
primates. *See* nonhuman primates
priming. *See* cultural priming
Pringle, C., 204
punctuation, 103
Pyers, Jennie, 60

Quiroga, R., 42

rabbits, 33–34
Railsback, Bruce, 202
Range, Friederike, 54
Ratcliff, K., 153
receivers. *See also* cultural agents; information
  associative learning, 13–14
  discrete representational knowledge, 16–21, 23, 25, 40–45
  diversity affects evolution, 7, 180–182, 188–191, 206–208
  innate knowledge, 14–15, 17
  Jablonka and Lamb, 12–13, 17–18
  must be prepared, 5, 6, 17–18, 32, 206
  must share a representational system, 67, 208–209, 212–214
  photocopier, 206
  Shannon-Weaver model, 12
  switch-like representations, 15–16, 19
Regenmortel, Marc Van, 197, 203–204

Regier, Terry, 79–80
Reis, Alexandra, 91
Renzulli, Joseph, 181–182, 188
representational medium, 68, 98, 99, 109. *See also* money
representations
  icon, index and symbol, 50
  increase reliability of inheritance, 20–21
  nonhuman, 50
  preadaptation for natural language, 50–51
Richerson, Peter
  cognitive attractors, 43–44
  conformity, 172
  discrete replicators not necessary for evolution, 40
  replicators not necessary for evolution, 45
Rodríguez Martínez, 109
Romney, A. K., 79
Rose, Andrew, 147–148, 156
Ross, Robert, 203
Rothschild, Michael, 195
Rotman, Brian, 135–136
Rusch, C. D., 79
Russon, Anne, 52
Ryan, R., 152

Salomon, Gavriel, 189
Schmandt-Besserat, Denise, 95–101, 121
Schofield, Jack, 193–194
Schumann, John, 59, 67, 71
Schwarz, Michael, 198
Seeger, Charles, 114–115, 117
Seyfarth, R., 14
Shannon, Claude, 12
sharecroppers. *See* discrete representations
Shashidhara, L. S., 227
Shaw, Ray, 63, 79, 94, 172, 178
Shennan, Stephen, 218, 227
Simmel, Georg, 145
Singh, Supriya, 137, 141, 142, 154
skills. *See* cultural ecology
Slegers, C., 142, 154
Sober, Elliott, 215
social brain hypothesis, 58
social status
  artefactual language use, 125, 164–165
  different groups, 161–165
  health, 159–165

nonhuman primates' health, 160–161
  subjective, 161, 164
  wealth, 159–160
spatial cognition, 81–82
species. *See* biological taxonomy; cultural taxonomy
speech. *See* natural language
Spelke, Elizabeth, 64
Spierdijk, Laura, 156, 159
spoked wheel, 195
Stanovich, Keith and Toplak, Maggie, 180
Stanovich, Keith and West, Richard
  Gricean communication, 76
  humans versus bees, 4
  myside bias, 179–180
  nonsocial interactions, 122
  two-system model of human reasoning, 173–176
status. *See* social status
Stengos, Thanasis, 156
Sterelny, Kim, 21
Stewart, Ian, 20
Stewart, Judith, 160
story telling, 104–106
swap space, 110–111
symbol, 50
synonyms, 66

taxonomy. *See* cultural taxonomy; biological taxonomy
technology
  Betamax and VHS, 193–194
  computing, 109
  education, 122–123
  evolution of, 192–196
  infrastructure, 194–195
  market forces, 196
  mobile phones, 192–193
  printing, 108–109
  prototypes, 195–196
  wheels, 195
Tehrani, Jamshid, 218–219, 227
text messaging, 102
The Selfish Meme, 72, 231
theory of mind, 60–61, 75
Tomasello, Michael, 53, 55, 59–60
Toplak, Maggie, 180
trade
  currency, 124
  currency effects, 146–148, 156

dependent on money, 128, 129, 131, 137
  dependent on trust, 143
  exchange media, 138
  reciprocal altruism, 148
transcription factors, 27
Truss, Lynn, 103
Turgenev, Ivan, 112

universal grammar, 70, 71, 72

Vellekoop, Michael, 156, 159
Verhaegen, Marc, 56
VHS. *See* technology
Viranyi, Zsófia, 54
virus
  characteristics, 186, 187
  complex evolution, 208
  DNA, 5, 222
  generations, 207
  taxonomy, 197, 203–204
Vohs, Kathleen, 152
Vygotsky, Lev, 112

Weaver, Warren, 12
Webb, John, 95
Weber, Ernest, 103
Webley, Paul. *See* Lea, Stephen and Webley, Paul
Weinzirl, John, 202
Werker, J., 72
West, Richard. *See* Stanovich, Keith and West, Richard
wheels, 195
Whiten, Andrew
  cultural taxonomy, 208
  human bias for hierarchical communication, 105–106
  human bias for social information, 104
  primate imitation, 51, 54
  primate theory of mind, 57
Whitman, William, 202
Wiebe, William, 202
Wiggins, Osborne, 198
Wilkie, J., 153
Wilkins, John, 205
Wilson, Mark, 160
Winford, Donald, 84–85
Wittgenstein, Ludwig, 197–199
Wood, J., 160

writing. *See also* clay tokens
  communication and representation,
    101–104
  divergence from mathematical notation,
    99–100
  early compositionality, 99
  early evolution, 98–100
  funerary function, 101
  metarepresentational, 102
  money (evolutionary analogy), 138–139
  proto-writing and social context, 95–96
  punctuation, 103
  significance, 90
  texts and emails, 102

Xing, Y., 28–29

Yair, Gad, 156
Yehuda, Rachel, 28
Young, H., 44

Zelizer, Viviana, 137, 151
Zellner, Debra, 161
Zuberbühler, K., 14